EcoProduction

Environmental Issues in Logistics and Manufacturing

Series Editor

Paulina Golinska-Dawson, Poznań, Poland

The EcoProduction Series is a forum for presenting emerging environmental issues in Logistics and Manufacturing. Its main objective is a multidisciplinary approach to link the scientific activities in various manufacturing and logistics fields with the sustainability research. It encompasses topical monographs and selected conference proceedings, authored or edited by leading experts as well as by promising young scientists. The Series aims to provide the impulse for new ideas by reporting on the state-of-the-art and motivating for the future development of sustainable manufacturing systems, environmentally conscious operations management and reverse or closed loop logistics.

It aims to bring together academic, industry and government personnel from various countries to present and discuss the challenges for implementation of sustainable policy in the field of production and logistics.

Adam Kolinski · Michał Adamczak
Editors

Digitalisation of the Greening Supply Chain

Springer

Editors
Adam Kolinski
Department of Engineering Sciences and Digital Technologies
Poznan School of Logistics
Poznań, Poland

Michał Adamczak
Department of Logistics
Poznan School of Logistics
Poznań, Poland

ISSN 2193-4614 ISSN 2193-4622 (electronic)
EcoProduction
ISBN 978-3-031-88917-2 ISBN 978-3-031-88918-9 (eBook)
https://doi.org/10.1007/978-3-031-88918-9

This Springer imprint is published by the registered company Springer Nature Switzerland AG
The registered company address is: Gewerbestrasse 11, 6330 Cham, Switzerland

Preface

The digitalisation of the greening supply chain is currently one of the most noticeable trends in the innovative development of supply chains. This results from the connection of issues such as digitalisation and the related changes in the generation and processing of data and the trend of reducing the impact of supply chains on the natural environment. These trends complement each other perfectly. Digital solutions allow for a significant reduction in the impact of the functioning of the supply chain on the natural environment. This impact can be observed both on a micro-scale (e.g. reducing paper consumption through the use of electronic documentation) and on a macro-scale (e.g. by providing information on the availability of products, their location and expiration date, which allows for better adjustment of their supply and reduction of the need to dispose of unsold goods). Thus, the development of digital technologies opens up completely new possibilities in building environmentally friendly supply chains.

The digitalisation of the greening supply chain also provides opportunities to improve the efficiency of the supply chain. For this reason, the key role in gaining a competitive advantage in the market is played by the use of new technologies and innovations, mainly in logistics processes. The use of innovative solutions in logistics processes is a very complex issue, both in terms of concept, as well as the practical implementation of the assumptions of the indicated innovations and supporting technologies. Gaining a competitive advantage in this area must be preceded by a detailed analysis of the needs for the digitisation of supply chain processes, the impact of implementing these solutions on the ecological aspect of the chain's functioning, and the possibilities of using this solution by partners cooperating with each other in the chain.

Actions that reduce the negative impact on the natural environment make more sense the wider the scope of entities in the supply chain (the more partners from the supply chain implement them and follow the outlined direction). Sectional activities (undertaken only by selected partners are also valuable but do not have such a large impact on a global scale). Digitising information flows provides opportunities to build solutions that are universal in nature thanks to the use of standards functioning in the digital world.

Digitalisation of the greening supply chain is a very complex issue that requires the separation of levels on which it will be considered. The scope of the presented monograph is divided into three main parts:

- Digitizing Information Flows in the Supply Chain,
- Use of IT Tools to Support Digitalisation and Optimalisation in Companies and Supply Chains,
- Use of Digital and Automative Technologies in Greening Supply Chains.

Through 11 chapters that positively passed at least two double-blind re-view processes, 31 authors from nine countries presents their latest scientific results providing significant contribution to the area of digitalisation of the greening supply chain.

The first part concerns digitizing information flow in the supply chain. It presents solutions in the field of: digital innovation in short food supply chains, using artificial intelligence in interdisciplinary risk management in road transport, reverse logistics in e-commerce and building sustainably in construction small enterprises. The range of industries and logistics functions in which digitalization can be used shows its universal nature. What's more, solutions from other industries can be an excellent inspiration for creating others dedicated to completely different conditions.

The second part focuses on the use of IT tools to support digitalization and optimization in companies and supply chains. The detailed topics discussed in this part concern: iterative simulations supported by GIS-class tools, stochastic programming in supply chain optimization and the impact of digitalization on ecological sustainability in warehousing. The issues discussed in this part focus on the optimization of the functioning of logistics processes. However, the authors present various technical solutions that allow for this optimization, using both programming methods, simulations and technical solutions functioning in warehouses in their research.

The third part of the monograph presents the use of digital and automatic technologies in greening supply chain. The authors of the individual chapters present: the role of green supply chain management in enterprise development, innovative supply chains of the circular economy, greening of e-commerce logistics by digitalizing and innovative analysis of surface layer parameters for sustainable product lifecycle management. And in this respect, the universality of digital solutions can be seen, which can also be used in detailed technical issues allowing for more effective management of the product lifecycle. Green supply chains are also those chains that provide products that can be used longer, which reduces the consumption of natural resources.

We would like to thank all reviewers whose names are not listed in the volume due to the confidentiality of the process. Their voluntary service and comments helped the authors to improve the quality of the manuscripts.

Poznań, Poland

Adam Kolinski
Michał Adamczak

Contents

Digitizing Information Flows in the Supply Chain

Embracing Digital Innovation in Short Food Supply Chains Through Alternative Food Networks

Maršenka Marksel, Piotr Nowak, David Strauß, and Tomislav Letnik

Abstract This chapter examines how digitalisation improves the efficiency and sustainability of short food supply chains (SFSCs). As agriculture evolves, it is transitioning towards digital practices and Agriculture 4.0, with a focus on improving food production efficiency. However, this evolution extends beyond food production, integrating processes at every stage of SFSCs, including storage, ordering, packaging, and delivery of local food. In these supply chains, Alternative Food Networks (AFNs) play a pivotal role in connecting various actors, fostering closer spatial-social relationships. The chapter examines the impact of digital technologies on logistics processes and the associated challenges in AFNs. Key findings indicate that digital tools improve the efficiency, competitiveness, and sustainability of AFNs, while offering recommendations for further innovations to enhance the resilience and effectiveness of local food systems.

Keywords Digitalisation · Agriculture 4.0 · Short Food supply Chain (SFSC) · Alternative Food Networks (AFNs)

M. Marksel (✉) · T. Letnik
University of Maribor, Maribor, Slovenia
e-mail: marsenka.marksel@um.si

T. Letnik
e-mail: tomislav.letnik@um.si

P. Nowak
Łukasiewicz Research Network, Poznań Institute of Technology, Poznan, Poland
e-mail: piotr.nowak@pit.lukasiewicz.gov.pl

D. Strauß
Fachhochschule des BFI Wien—Hochschule für Wirtschaft, Management and Finance, University of Applied Sciences BFI Vienna, Vienna, Austria
e-mail: david.strauss@fh-vie.ac.at

A. Kolinski and M. Adamczak (eds.), *Digitalisation of the Greening Supply Chain*, EcoProduction, https://doi.org/10.1007/978-3-031-88918-9_1

1 Introduction

The global challenges of climate change, economic crises, pandemics, and geopolitical instability have contributed to the vulnerability of various systems around the globe (Desalegn et al. 2022; Buhaug and Von Uexkull 2024). These disruptions highlight the need for more resilient, sustainable practices, particularly in the food supply chain, which plays a crucial role in ensuring food safety and environmental stability (Mcclements et al. 2020; García-Díez et al. 2021). With increasing emphasis on sustainability, local food systems and short food supply chains (SFSCs) are being recognized as very promising solutions (Jia et al. 2024). These concepts emphasize regional food production, processing, and distribution, which can reduce dependence on long, complex and vulnerable global supply chains (Chi Ffoleau and Dourian 2020; Baldwin and Freeman 2022). Localized systems offer substantial environmental, economic, and social benefits by minimizing transportation distances, lowering carbon emissions, and fostering stronger connections between producers and consumers (Borsellino et al. 2020; Liu et al. 2023).

However, while SFSCs centre on logistics and proximity, alternative food networks (AFNs) take a broader, value-driven approach by promoting sustainability, ethics, and systemic change. Together, SFSCs and AFNs strengthen local self-sufficiency, reduce reliance on global markets, and boost resilience to external shocks.

Despite the advantages, transitioning to these systems introduces various challenges. Some of the broader issues include the difficulty of scaling local systems to meet growing population demands, limited infrastructure for processing and distribution, and regulatory barriers that often favour larger, centralized operations (Berti and Mulligan 2016). Policy efforts often fall short to fully understand and address the unique demands of local food systems, leaving many small-scale initiatives struggling to compete with industrial food supply chains (Kapała 2022; Stein and Santini 2022).

Significant obstacles worth to consider are related to logistics. Challenges arise due to the geographic dispersion of small-scale producers, the seasonal nature of production, and the complexities of handling small, varied quantities of perishable goods (Mittal et al. 2018). Coordinating transportation from multiple locations increases costs and strains cold chain logistics, especially with smaller, varied shipments (Govindan et al. 2014). Packaging requirements differ widely across products, adding complexity to storage and transport, while small-scale producers often lack access to advanced warehousing and transportation infrastructure (Barbosa-Cánovas et al. 2011). Ensuring traceability and maintaining food safety can be challenging in such a fragmented system, with cooperation among diverse stakeholders often constrained by differences in available resources (Bosona and Gebresenbet 2013). These factors can lead to inefficiencies in transportation, storage, and order management and hinder development of AFNs.

Digitalization presents a promising solution to address logistics inefficiencies while improving the overall performance of SFSCs. Agriculture has in general already embraced digital innovation through concepts like "Agriculture 4.0" and the "digital agricultural revolution," but in order to take the effect, this potential must extend beyond the farm production, encompassing the entire supply chain (Lezoche et al. 2020). Digital tools such as inventory management systems, e-commerce platforms, and route optimization software can greatly improve logistics in SFSCs, by streamlining transportation, cutting fuel consumption and emissions, and lowering costs (Todorovic et al. 2018). Even though digital solutions for logistics operations are available, they are seldom implemented in SFSCs. This limited adoption raises important questions about the challenges and opportunities for integrating digital solutions effectively within these networks.

Considering these challenges, our study aims to clarify the defining characteristics of Alternative Food Networks (AFNs), their role within Short Food Supply Chains (SFSCs), and the specific logistics issues they face. This clarification is urgently needed because various definitions and interpretations of AFNs exist, leading to confusion and inefficiencies in addressing their logistics needs. By establishing a clear framework and understanding of AFNs, we aim to facilitate the development of targeted strategies that can effectively address their unique challenges, ultimately contributing to the support further development of local food systems.

2 Research Methodology

This study aims to understand the potential influence of digital innovation on Short Food Supply Chains (SFSCs) and Alternative Food Networks (AFNs), as well as to identify opportunities for improving logistics, efficiency, and sustainability. To achieve these objectives, a structured methodology is employed that integrates a literature review with an analysis of diverse case studies, allowing for both theoretical insights and practical applications to be considered. The literature review focuses on synthesizing relevant academic studies, industry reports, and policy documents published between 2009 and 2024.

This period reflects significant advancements in agriculture, logistics, and supply chains driven by the need for innovation and resilience following the 2008 global economic crisis (Hausman and Johnston 2014). Key technological milestones, including blockchain, IoT, and AI, emerged during this time, transforming food networks by enhancing traceability, efficiency, and sustainability (Osato Itohan Oriekhoe et al. 2024). Policy shifts and global initiatives, such as the Sustainable Development Goals, European Green Deal, Farm to Fork strategy, further supported the growth of local food systems and circular economies, reinforcing the role of AFNs and SFSCs (Commission 2024a, 2024b; Skalkos 2024). Additionally, changing consumer behaviour and crises like COVID-19 highlighted the demand for transparent, sustainable, and decentralised food networks, making this period crucial for understanding their development (Atalan-Helicke and Abiral 2021; Szabo et al. 2021; Desalegn et al. 2022). By selecting articles from 2009 to 2024, the literature review effectively captures the intersection of technological advancement, policy evolution, and shifting market dynamics that shape the current landscape of SFSCs and AFNs.

Initial literature review was conducted using the Web of Science database considering the following key topics: digital technologies in agriculture (230 articles), short food supply chains (114 articles), alternative food networks (100 articles), and logistics innovations in the food industry (300 articles). In the second phase only the most relevant works, selected based on their alignment with the research objectives, were included in the in-depth analysis. The review specifically addresses areas like inventory management systems, route optimization, and collaborative digital solutions to logistics challenges. It also examines the barriers and enablers of digital adoption within AFNs, revealing factors that either facilitate or hinder the successful integration of digital tools.

Our analysis offers AFNs a foundation for understanding and implementing digital solutions in daily operations, with the goal of enhancing operational efficiency. Furthermore, this review aims to address a gap in the existing literature, providing comprehensive insights and suggesting future directions to advance adoption of digital logistics solutions to achieve sustainable short food supply chains.

In continuation, *Chapter 3* discusses the role of Alternative Food Networks (AFNs) in enhancing local collaboration and sustainability, highlighting various organizational forms and their impact on local food systems. *Chapter 4* highlights the impact of digitalization and new technologies on logistics efficiency, delving into specific digital solutions appropriate for SFSCs. *Chapter 5* discusses the key findings and strategic recommendations, including effective strategies and innovative practices that can be scaled or adapted to similar contexts, with a brief conclusion presented in *Chapter 6*. Together, these chapters provide a comprehensive exploration of the opportunities and challenges faced by AFNs and SFSCs, guiding stakeholders on leveraging digital tools to enhance local food system resilience and efficiency.

3 Alternative Food Networks (AFNs) and Their Role in Short Food Supply Chains (SFSCs)

This chapter explores the broader role of AFNs within Short Food Supply Chains (SFSCs). It examines how AFNs contribute to local collaboration, sustainability, and food sovereignty, addressing the challenges posed by industrialized food systems. In doing so, AFNs serve as a potential model for making food supply chains shorter, more transparent, and better aligned with community values. The detailed forms and characteristics of AFNs, as well as the various ways they operate, will be elaborated upon in the following sections, shedding light on how these networks can help reshape our food systems.

3.1 Understanding Alternative Food Networks (AFNs)

Alternative Food Networks (AFNs) are typically seen as networks that ***help bridge the gap between producers and consumers***, reducing both geographical and social distances between them (Gori and Castellini 2023). These networks aim to create more sustainable, ethical, and localized food systems by creating direct connections between producers and consumers (Reckinger 2022). They remain a vital part of the broader movement towards more ethical and sustainable food systems (Cerrada-Serra et al. 2018).

Although the term may seem unfamiliar to some, AFNs have been around for quite some time. They have emerged in the 1990s as a response to the industrialization and globalization of the food system. In Europe, AFNs were influenced by environmental and organic agriculture movements and reforms in the European Union's Common Agricultural Policy (CAP), which shifted towards more decentralized and multifunctional agriculture. Health and food safety concerns, such as the outbreak of Bovine Spongiform Encephalopathy (BSE) in 2001 or Covid-19 in 2020, also played a significant role in driving consumer demand for better food quality and transparency. AFNs continue to evolve, addressing criticisms related to exclusivity, elitism, and the impact of capitalist markets on their goals (Edwards 2016). As AFNs have evolved, a wide range of agricultural initiatives within unconventional food markets has also emerged (Brinkley 2018; Michel-Villarreal et al. 2019). This expanding diversity has made it difficult to define all the initiatives that fall under the broader category of AFNs with a single, unified definition (Bazzani and Canavari 2013; Barbera and Dagnes 2016).

Question. Given the numerous concepts associated with AFNs, a key question emerges: what are the ***criteria*** that definitively ***determine*** whether an initiative qualifies as an ***AFN***?

To address this question, we can rely on the definition of primary and secondary characteristics of AFNs. ***Primary characteristics*** are the core, essential features that define AFNs and distinguish them from conventional food systems. They are present in almost all AFN forms, without these, a network would likely not be considered an AFN. ***Secondary characteristics*** often enhance the AFN business model but aren't urgently needed to function as AFN. In other words, an AFN can exist without some of these secondary characteristics, but these features often contribute to making the network more distinct or aligned with specific goals. Main primary and secondary characteristics of AFNs are listed in Table 1.

According to Tregear (2011) the ***primary characteristics*** of AFNs include ***shorter distances*** between producers and consumers, both geographically and in terms of reduced intermediaries. There is a strong emphasis on ***local production***, where the majority of products are locally sourced. AFNs operate within a ***commercial framework***, functioning within a market-based economy, and aim to ***bridge the gap between rural food producers and urban consumers***. They also involve specific ***food purchasing venues***, such as food cooperatives, farmers' markets, and Community Supported Agriculture (CSA) initiatives. Additionally, AFNs are committed to

Table 1 Primary and secondary characteristics of AFNs

Primary	Secondary
Shorter distances between producers and consumers	Reduced packaging aligned with sustainability goals
Focus on local production for nearby sourcing	Small farm size emphasizing traditional, localized agriculture
Market-based (commercial) operation	Links to fair trade and perception of quality
Connecting urban customers with rural producers	Small farm size and scale
Arranging food purchasing venues	Promotion of organic farming
Embedding social, economic, and environmental sustainability values	Aesthetic and sensory food qualities

Source Own elaboration based on (Tregear 2011; Kneafsey et al. 2013)

the ***social, economic, and environmental aspects*** of sustainable food production, distribution, and consumption.

According to Kneafsey et al. (2013) the ***secondary characteristics*** of AFNs include factors like ***small farm size***, the use of ***organic farming*** methods, and a contrast to large-scale industrial agribusiness. ***Social integration*** plays a crucial role, encompassing elements such as reputation, personal trust, information sharing, cooperation, and social capital. ***Ecological embeddedness*** is also important, involving a strong personal connection to the place of production, active participation in shaping ecological conditions, and sharing natural and geographical knowledge between producers and consumers. AFNs often maintain ***links*** with broader movements, like Slow Food for local connections, organic certification for ecological ties, or Fair Trade for social equity. The ***perception of quality*** in these networks tends to focus on the image of traditional producers rather than strictly on environmental performance, except in the case of direct marketing of organic food, where effective practices have been demonstrated AFNs also prioritise ***reduced packaging*** to minimise waste and align with sustainability goals, often adopting eco-friendly solutions. Being part of initiatives or platforms, such as certification programmes like BioAustria, is common (Austria 2024). Finally, AFNs value the ***aesthetic and sensory attributes*** of food, where the pleasure of taste and smell plays a significant role in consumer evaluation of food quality.

Question. How do the **various organizational forms** within Alternative Food Networks (AFNs) embody and adapt the primary and secondary characteristics of AFNs to address diverse community needs and operational contexts?

While understanding the primary and secondary characteristics, it's also important to recognise the various organisational forms these networks can take. These forms reflect the diversity of AFNs in practice, with each adapting the core principles to various contexts and structures. The most common forms of AFNs include ***producer cooperatives****, where* farmers or producers join forces to collectively manage and market their goods, with ownership and operation by the producers themselves to achieve common economic goals (Cerrada-Serra et al. 2018). ***Platforms*** serve as digital or physical infrastructures that connect producers and consumers, streamline the distribution process, and support network functioning. They are facilitating the exchange of information, goods, and services among farmers, suppliers, distributors, consumers, and other participants in the AFN sector (Goszczynski et al. 2019). ***Producer-consumer partnerships***, such as community-supported agriculture (CSA), establish collaborative relationships between food producers and consumers. Their main aim is to create direct, sustainable, and mutually beneficial connections, often bypassing traditional intermediaries, ensuring fair prices, and providing high-quality, locally sourced food (Martindale 2021). ***Food cooperatives***, which are member-owned and operated, provide an alternative to conventional food retail by emphasizing local, sustainable, and ethical food production and distribution, serving the interests of members who are typically both owners and consumers (Cerrada-Serra et al. 2018). ***Direct retail initiatives*** allow producers to sell directly to consumers, bypassing traditional intermediaries to strengthen producer-consumer connections, enhance transparency, and support local economies (Ashtab et al. 2020). ***Agro-ecotourism***, represented by farm shops, attracts visitors to agricultural areas to experience and learn about farming practices, local food production, and rural life (Pinna 2016). Lastly, "***consumers as producers***" initiatives, such as community gardens, involve community members collectively managing and cultivating shared garden spaces, emphasizing local food production, community engagement, and sustainable practices (Opitz et al. 2017).

Question: How do **varying levels of complexity** within Alternative Food Networks (AFNs) influence their operational efficiency, sustainability, and ability to meet diverse consumer and producer needs?

AFNs can range from simple, direct exchanges between producers and consumers to solutions including intermediaries, to more advanced, highly organized systems supported by advanced digital technology (Laginová et al. 2023). ***Direct AFNs*** are basic, informal networks with direct relationships between producers and consumers. Producers, like small-scale farmers, sell directly to consumers, they often manage deliveries themselves, either through home delivery or at farmer's markets. This approach requires minimal transportation, packaging, or storage facilities. ***Intermediary AFNs*** are more advanced forms, which involve intermediaries who manage tasks such as collection, packaging, and distribution. This structure adds a layer of professionalism to the process. Intermediaries—like cooperatives or food hubs—collect products from local producers. They sort and package goods to meet market demands, ensuring quality standards. Distribution to retail outlets, restaurants, or consumers is organized through small-scale logistics networks, often using dedicated delivery vehicles and sometimes small warehouses or cold storage facilities. ***Advanced AFNs*** are the most complex and efficient form. They resemble conventional retail chains, with integrated IT systems and branded store networks. Advanced AFNs use supply chain management software to track inventory, sales, and delivery schedules in real-time. Products from multiple suppliers are processed and stored in central warehouses before distribution. Goods reach consumers through branded store networks or online platforms, creating a consistent shopping experience. These systems often involve third-party logistics (3PL) providers, refrigerated trucks, and centralized hubs to ensure timely, efficient deliveries.

> **Important** From understanding the specifics of each complexity level, we can summarise that direct AFNs typically have minimal or no digitalisation in their logistics processes, intermediary AFNs operate with a medium level, and advanced AFNs demonstrate high levels of digitalisation. This alignment between digitalisation and organisational complexity highlights how each type of AFN, whether direct, intermediary, or advanced, integrates digital tools to enhance logistics efficiency and address network-specific needs.

3.2 *What is Short Food Supply Chain (SFSC)?*

While AFNs encompass a wide range of alternative systems aimed at transforming food production, distribution, and consumption with a focus on values and ethics, Short Food Supply Chains (SFSCs) specifically aim to shorten the supply chain, making it more direct. In essence, SFSCs can be understood as a distribution model that minimises the number of intermediaries between local food producers and consumers. These often involves direct sales from farmers to consumers, local markets, or community-supported agriculture (CSA) schemes. SFSCs emphasise the efficiency and transparency of the food chain itself, with a priority on reducing intermediaries and often on local sourcing to ensure fresh, traceable products. Key characteristics of short food supply chains are summarized in Table 2 (Parliament 2016).

The aforementioned characteristics make short supply chains an appealing alternative to conventional, longer supply chains, particularly for those with an interest in sustainability and in supporting local economies (Kneafsey et al. 2013). The main logistics processes involved in SFSC are procurement, storage, inventory management, order picking and packing, transport and delivery and information management related to the local food products.

Question: How do this **key logistics processes impact** the efficiency, sustainability, and consumer satisfaction within Short Food Supply Chains (SFSCs)?

Procurement involves sourcing raw materials or products directly from local farmers or producers. The focus is on minimizing the distance between the source and the consumer to ensure freshness and reduce transportation costs (Dominic 2023). Key activities include identifying local suppliers, negotiating terms, and ensuring quality standards. ***Storage*** is crucial to maintain the quality and freshness of food products. In SFSCs, storage facilities are often located close to the production sites or distribution points. In that case, key activities include temperature control, inventory management, and ensuring hygienic conditions. Efficient ***inventory management***

Table 2 Main characteristics of short food supply chains (SFSC)

Local focus	Direct sales
Products are typically sourced and sold within a specific geographic area, promoting local economies	Producers sell their goods directly to consumers, reducing the number of intermediaries
Transparency	Sustainability
Consumers have better knowledge of where their food comes from and how it is produced	SSCs often emphasize environmentally friendly practices, such as reduced transportation distances, which lower carbon footprints
Community engagement	Fair pricing
SSCs can strengthen social ties within communities by fostering direct relationships between producers and consumers	Farmers often receive a fairer price for their products compared to conventional supply chains

Source Own elaboration based on (Parliament 2016)

helps in tracking stock levels, reducing waste, and ensuring that products are available when needed. This includes monitoring stock levels, forecasting demand, and managing replenishment cycles. The next important logistic process is ***order picking and packing***. This process involves selecting the right products from inventory to fulfil customer orders and packing them appropriately for delivery (Mecalux 2023). Key activities are related to order verification, product selection, and packaging. ***Transport and delivery*** are mainly focused on minimizing the distance travelled and ensuring timely delivery to maintain product freshness. It comprises of route planning, vehicle management, and delivery scheduling. Last but not least, logistic process in SFSC is very much dependent on efficient ***information management***. Effective communication and information flow between producers, distributors, and consumers are essential for the smooth operation of SFSCs. Key activities here are using digital tools for tracking and managing orders, maintaining transparency, and facilitating direct communication (Dominic 2023).

Important Advanced procurement, storage, inventory management, order picking, transport and information management are designed to enhance the efficiency, sustainability, and transparency of short food supply chains, making them a viable alternative to traditional, longer supply chains.

3.3 Role of Alternative Food Networks (AFNs) in Short Food Supply Chains (SFSCs)

Alternative Food Networks (AFNs) play a crucial role in Short Food Supply Chains (SFSCs) by promoting sustainable, local, and direct food distribution systems. These roles make AFNs a vital component of SFSCs, contributing to more resilient, sustainable, and transparent food systems.

Question: How do Alternative Food Networks **(AFNs) contribute** to local economic resilience, environmental sustainability, and community engagement, while enhancing transparency and fairness in the food supply chain?

AFNs support local farmers and producers by providing them with direct access to markets, thereby ***supporting local economies.*** They are reducing the distance food travels from farm to fork, and so helping lower carbon emissions and ***promote environmentally friendly practices***. AFNs also foster ***transparency*** in the food supply chain, allowing consumers to know where their food comes from and how it is produced. Direct sales through AFNs ensure that consumers receive ***fresher and higher-quality products*** compared to those in longer supply chains. AFNs often involve community-supported agriculture (CSA) schemes and farmers' markets, which ***strengthen social bonds within communities***. And lastly, by eliminating intermediaries, AFNs can offer ***fairer prices*** to both producers and consumers (Petruzzelli et al. 2023).

Question: What **challenges** do Alternative Food Networks (AFNs) encounter, particularly as they strive to maintain their foundational values?

AFNs face several ***challenges*** as they strive to create more sustainable and ethical food systems. AFNs often ***struggle to scale up*** while maintaining their core values of sustainability and localism. As they grow, they often face logistics challenges in distribution and struggle to maintaining direct relationships between producers and consumers (Kondo and Suzuki 2022). They also struggle with ***economic viability,*** meaning balancing affordability for consumers with fair compensation for producers. Ensuring that local, organic, or sustainably produced food remains accessible without compromising the incomes of small-scale farmers can be difficult (Atalan-Helicke and Abiral 2021). In some regions, particularly where ***social capital*** is low, building trust and cooperation among stakeholders can be challenging. This is crucial for the success of community-supported agriculture and other AFN models (Poças Ribeiro et al. 2021).

Furthermore, navigating complex ***regulatory environments and barriers*** can be intimidating for small-scale producers and AFNs. Regulations designed for large-scale industrial food systems may not be suitable for smaller, localized operations (Poças Ribeiro et al. 2021). ***Competing on the market*** with large-scale industrial food systems that benefit from economies of scale and extensive distribution networks can be tough (Kondo and Suzuki 2022). ***Rasing consumer awareness and engagement*** about the benefits of AFNs and encouraging them to change their purchasing habits requires ongoing effort. Overcoming the convenience and lower prices of conventional food systems is a persistent challenge of AFNs (Kondo and Suzuki 2022).

Question: How can Alternative Food Networks (AFNs) overcome the challenges they face while staying true to their core values?

Despite these challenges, Alternative Food Networks (AFNs) continue to ***innovate and adapt***, consistently seeking creative methods to strengthen connections between producers and consumers while promoting sustainable food practices. Based on AFN findings, key advantages can be nowadays achieved mainly through digitalization. Majority of AFNs take the opportunity provided with establishment of digital platforms. ***Online marketplace*** is one example of that, where producers can sell directly to consumers, which can help AFNs scale up and reach a broader audience. These platforms can facilitate transactions, manage orders, and provide a space for customer reviews and feedback (Michel-Villarreal et al. 2021). Implementing digital tools for ***supply chain management*** can help with tracking and managing inventory and improving the efficiency and transparency of AFNs. Technologies like blockchain can ensure traceability and authenticity of products (Forno et al. 2024). Using ***data analytics*** helps to understand consumer preferences, optimize production, and forecast demand that can support AFNs making informed decisions. This can lead to better resource management and reduced waste (Michel-Villarreal et al. 2021). Manny AFNs are using ***social media platforms*** for marketing and community engagement, which can increase visibility and attract more consumers. It also provides a way to educate the public about the benefits of sustainable and local food systems (Michel-Villarreal et al., 2021). The other option is to take advantage of ***mobile applications*** for consumers to easily find and purchase products from AFNs enhancing convenience and accessibility. Mobile applications can also provide information on product origins, nutritional content, and recipes (Forno et al. 2024). E-learning and education in the form of ***online courses and resources*** for farmers and producers can help them adopt sustainable practices and improve their business skills. This can include training on digital tools, marketing strategies, and sustainable farming techniques (Michel-Villarreal et al. 2021). Last but not least, ***digital community platforms*** are available, which can facilitate the creation of online communities where consumers and producers can share experiences, knowledge, and support. This can strengthen the social capital that is crucial for the success of AFNs (Forno et al. 2024).

Table 3 outlines the various challenges faced by Alternative AFNs and illustrates how these challenges can be transformed into opportunities by digital technologies. By embracing these digital tools, AFNs can enhance their resilience, efficiency, and ability to meet the growing demand for sustainable and ethical food options. While many digital technologies offer a range of solutions, the tools specifically aimed at improving logistics efficiency are crucial, particularly in the context of SFSCs. These digital tools are further explored in the following chapter.

Alternative Food Networks (AFNs) provide a viable pathway to rethinking food supply systems by addressing many of the challenges inherent in large-scale, industrial food networks. Despite their potential, AFNs face significant challenges related

Table 3 AFNs challenges and opportunities via digitalisation

Challenges	Opportunities (digital technologies)	Explanation
Scalability	Online marketplaces, supply chain management	Digital platforms and tools help scale AFNs without compromising core values, by improving distribution and customer engagement
Economic visibility	Data analytics, mobile applications	Data analytics helps optimize production and reduce waste, while mobile apps increase sales and lower transaction costs
Social capital	Community building, social media and marketing	Online communities and social media foster trust and cooperation, crucial for building relationships among stakeholders
Regulatory barriers	Supply chain management, e-learning and training	Supply chain tools ensure traceability and compliance with regulations, while e-learning helps producers understand complex rules and legalization
Market competition	Social media and marketing, online marketplaces	Digital platforms provide more visibility and direct access to consumers, helping AFNs compete with larger systems
Consumer awareness and engagement	Social media and marketing, mobile applications	Social media and apps educate and engage consumers, fostering awareness of AFN benefits and making sustainable choices more accessible

Source Own elaboration based on: (Michel-Villarreal et al. 2021; Forno et al. 2024)

to scalability, regulatory constraints, market competition, and consumer awareness and mainly logistics. However, these challenges also present opportunities for innovation, particularly through the adoption of digital solutions.

Important AFNs strengthen Short Food Supply Chains (SFSCs) by boosting local economies, reducing environmental impact, and fostering direct producer–consumer connections. Key improvements include collaborative logistics, digital tools for efficiency and transparency, and supportive policies for small producers. These strategies enhance AFN sustainability, scalability, and consumer engagement.

4 Digital Solutions for SFSC Logistics Challenges of AFNs

In the previous chapters, numerous challenges faced by stakeholders in Alternative Food Networks (AFNs) have been identified and discussed. This chapter aims to systematically address these challenges by proposing digital solutions in logistics processes that stakeholders can utilize to improve their operations. These solutions can be in general categorized into two main frameworks: those designed for individual stakeholders and those that facilitate collaborative logistics. By examining various logistics aspects, a comprehensive understanding can be gained of how digitalisation addresses logistics challenges in SFSCs within the broader context of AFNs. While there are numerous potential tools available, it is essential to focus on those that are most suitable for addressing specific problem areas. It is not our intention to present an exhaustive list of solutions; rather, we offer a selection of highly appropriate options for each category. Before presenting the solutions, some general insights of digitalization and its influence on agriculture are presented.

4.1 Understanding Relevant Digitalization Aspects

The advent of Industry 4.0 has introduced advanced technologies such as Internet of Things (IoT), blockchain and artificial intelligence (AI). All of these technologies are set to revolutionize the logistics sector (Cao et al. 2022; Albrecht et al. 2024).

Question: What **characteristics** must digital tools have to effectively support the evolving needs of food networks?

New digital tools must be capable of functioning in a **seamless** manner alongside existing systems, in order to prevent the formation of information silos (Lezoche et al. 2020). Achieving seamless communication between all systems is essential to realizing the goal of a successful digital transformation.

Another crucial aspect is ***scalability***. As food networks expand, the technologies that support them must also evolve. Digital solutions that are unable to adapt to changing business needs can impede growth. Therefore, scalable and flexible solutions that can accommodate future growth are essential for long-term success.

Furthermore, the ability ***to share information in real time*** is of paramount importance. To illustrate, Internet of Things (IoT) devices can monitor conditions during transportation, while blockchain technology ensures transparency throughout the supply chain. Access to real-time data allows companies to make informed decisions and adjust processes in a timely and efficient manner, particularly when managing perishable goods (Albrecht et al. 2024). A social media account became an indispensable digital instrument for communication, marketing, and engagement. It permits companies to advertise their products, disseminate their narratives and establish connections with a broader audience. Social media platforms such as Instagram, Facebook, and Twitter empower producers to interact with consumers and stakeholders through regular posts, videos, and real-time updates (Bos and Owen 2016; Viciunaite 2023).

Question: What is the **influence of digitalization** on the agricultural sector, particularly in transforming practices and optimizing short food supply chains?

The influence of ***digitalization*** on the agricultural sector is considerable. The advent of modern farming machinery has seen the introduction of numerous digital tools and optimizations, which have the potential to transform agricultural practices

(Lezoche et al. 2020). A significant advancement in this field is the utilization of decision support systems, which employ sophisticated technologies such as sensors and drones. These systems gather crucial data on a range of agricultural aspects, including weather conditions, geographical factors, and the behavior of animals and crops, offering real-time insights into the entirety of the farm life cycle (Mondino and Gonzalez-Andujar 2019). In the context of short food supply chains, digital solutions have been discussed and implemented across several key areas, including warehousing, ordering, transportation, packaging, reverse logistics and communication. The influence of digital advancements on these areas is evident, nevertheless, the level of implementation in Short Food Supply Chains is far from optimal.

Important Effective digital transformation in AFNs requires tools that integrate seamlessly with existing systems, are scalable, and support real-time data sharing. Technologies like IoT and blockchain boost transparency, efficiency, and stakeholder engagement, helping AFNs meet consumer expectations and maintain product quality.

4.2 Solutions for Individual Stakeholders in Short Food Supply Chains

Introduction of technological solutions is in general initiated based on identified specific challenges of AFN stakeholders. By integrating these digital solutions, stakeholders not only enhance their operational capabilities but consequently also contribute to the overall effectiveness of the supply chain.

Question: What are the specific **logistics challenges** faced by AFN stakeholders, and what are the most **promising digital solutions** or logistics concepts available to address them?

A common logistics challenge for AFNs worth mention is selection of the most suitable means of transportation, particularly to meet the demands of cold chain logistics (Behdani et al. 2019; Görçün et al. 2024). For that purpose, AFNs can take advantage of ***fleet management tools*** which enables them to select vehicles most appropriate for their specific delivery needs (Zhong et al. 2017; Fares et al. 2024). The use of most appropriate vehicles not only preserves product quality but also enhances fuel efficiency and reduces emissions, which is particularly important for small, localized supply chains that prioritize sustainability (Inkinen and Hämäläinen 2020). In urban areas, local producers have adopted electric vans for distribution, reducing emissions and supporting their environmental goals (Cagliano et al. 2017). Frequent, small-scale deliveries, which are characteristic for SFSCs often face inefficiencies due to traffic congestion and suboptimal routing. This leads to increased delivery times, fuel costs, and potential product spoilage (Miwa and Bell 2017; El Yaagoubi et al. 2024). To mitigate these issues, ***route optimization tools*** can be used, which take advantage of real-time traffic data and predictive analytics to enhance delivery efficiency (Mahendran et al. 2024). Route optimization can be used for various AFN delivery services, such as communal home meal delivery, to identify the fastest routes while avoiding congested areas, thereby reducing delivery times, fuel costs, and emissions (Bräysy et al. 2009). These tools are especially beneficial during high-traffic periods and when managing many delivery points, where they can demonstrate their full advantages.

Maintaining optimal stock levels presents a significant challenge for Alternative Food Networks (AFNs) due to fluctuating supply and demand, often influenced by the seasonality of products. These fluctuations can lead to waste from over-stocking or missed sales opportunities from stockouts, especially for perishable and seasonal goods. Employing ***inventory optimization tools*** that offer real-time tracking and demand forecasting can mitigate these issues (Singh et al. 2024). This kind of tools are particularly valuable for small-scale producers of seasonal products, which need to maximize yield during harvest while minimizing potential losses (Woolley et al. 2020). Alternative Food Networks are dealing also with the issue of managing temperature-sensitive products. This requires constant monitoring of storage conditions, which can be resource-intensive and prone to human error, leading to compromised product quality. ***Warehouse Management Systems (WMS)*** equipped with IoT sensors address this challenge by automating stock handling and tracking storage conditions (Mittal et al. 2018; Jagtap et al. 2021).

This automation allows producers to maintain high standards and streamline their operations, which is a crucial benefit for small and medium-sized businesses that lack extensive logistics resources(Bastard and Chaillet 2023). Small businesses often lack the capital or resources to develop in-house logistics infrastructure and services, hindering their ability to scale operations and meet demand fluctuations. ***Digital logistics platforms*** provide a solution by enabling AFN stakeholders to outsource logistics services as needed, offering a flexible approach for small businesses (Akac et al. 2021). For instance, a family-owned farm might rely on a third-party logistics provider to handle peak season deliveries, accessing reliable logistics support without the overhead costs of owning and maintaining a fleet. This outsourcing flexibility is particularly advantageous for seasonal producers, as it allows them to scale operations in response to demand without significant investment (Arias-Aranda et al. 2011).

Important Individual AFN stakeholders can tackle logistics challenges by adopting digital solutions like fleet management, route optimization, inventory management, and flexible logistics platforms. These tools improve efficiency, cut costs and emissions, and enable small producers to scale operations in line with demand.

4.3 Supply Chain-Level Solutions Enabling Collaboration

In the context of collaborative logistics, lack of coordination among multiple stakeholders can lead to inefficiencies and higher operational costs due to underutilized logistics capacities. Digital solutions promote cooperation by enabling resource-sharing and more efficient logistics operations across AFNs and SFSCs.

Question: What are the **common logistics challenges** in AFNs, and how can **collaborative solutions** address these issues to enhance sustainability and efficiency?

For instance, cooperatives can use shared warehousing facilities to store products in centralized locations, reducing individual storage costs and improving product preservation (Rogers et al. 2020). Individual AFN stakeholders may have underutilized logistics capacities. Without coordination, this leads to inefficiencies. ***Resource-sharing platforms***, such as those for shared warehousing and transportation, optimize logistics by enabling stakeholders to pool capacities based on current needs (Furtado et al. 2014). By leveraging shared resources, digital platforms allow small producers to access facilities that might otherwise be financially too expensive, enhancing the sustainability and efficiency of AFNs.

In addition to that, small producers also face fluctuating storage needs due to seasonal variations, making long-term warehouse leasing financially unfeasible and leading to storage shortages or underutilized space. ***Warehouse-as-a-Service platforms*** (WaaS) offer on-demand storage solutions, allowing local producers to scale storage capacity in response to seasonal needs (Xenou et al. 2021). This flexible approach is essential for small producers in short food supply chains, where storage needs fluctuate, and long-term commitments are financially infeasible. When WaaS allows multiple producers to share storage facilities, it contributes to collaborative logistics by enabling joint product storage, which can lead to reduced costs and logistics synergies. Various and single-use packaging is another problem of AFNs, which contributes to waste and inconsistency in logistics operations. Unification and standardization of ***packaging equipped with digital tracking*** addresses this issue (Parrag et al. 2022). The integration of technologies like intelligent food packages and wireless sensor networks can enhance monitoring capabilities, improve logistics efficiency, reduce food loss, which bring benefits to all AFN members (Vanderroost et al. 2017). Moreover, it also has a broader collaborative impact in case of creating a standardised reusable packaging ecosystem, allowing multiple AFN members to share packaging resources.

Several potentials are available in case of addressing the problem of redundant transport operations, which arise due to uncoordinated shipments among AFN members. ***Joint transport ordering and delivery systems*** enable producers to consolidate shipments, minimizing redundant trips and transportation costs (Fikar and Leithner 2021). In regions with overlapping delivery routes, such as urban centres, multiple producers can coordinate deliveries through a shared platform, reducing environmental impact and optimizing logistics (Nordmark et al. 2012). Last-mile delivery challenges are crucial issue, because it increases costs and environmental impact while also distancing consumers from producers. ***Systems for coordinated***

pickup points is a solution, that provides direct connection between producers and consumers by establishing centralized collection locations (Bosona et al. 2011). Food hubs serve as innovative organizational arrangements that bridge gaps between small producers and consumers, helping to create value-based food supply chains and improve market access for small farms operations (Berti and Mulligan 2016). This setup strengthens community connections and promotes sustainable consumption, as consumers engage directly with local producers.

Inefficiencies, bottlenecks, and delays in the supply chain often result from a lack of accurate demand forecasting and resource allocation. ***Supply chain optimization tools*** equipped with predictive analytics enhance the overall flow of goods by allowing stakeholders to better anticipate demand and allocate resources accordingly (Ibiyemi and Olutimehin 2024). AFNs could utilize optimization tools to synchronize production with market needs, ensuring timely deliveries and minimizing supply chain delays (Esteso et al. 2022). These tools could support efficiency by reducing bottlenecks and enabling a smoother, more reliable supply chain.

Important Collaborative digital solutions enhance logistics efficiency and sustainability in AFNs by enabling resource-sharing, delivery consolidation, and coordinated pickup points. Shared warehousing, reusable packaging tracking, and demand forecasting help minimize costs, reduce waste, and strengthen community connections.

5 Discussion and Strategic Recommendations

AFNs face several logistics challenges, including managing small-scale, geographically dispersed producers, seasonal variations, and the complexities of handling perishable goods. These challenges make it difficult to scale local systems, maintain cold chains, and coordinate transport from multiple locations. Limited infrastructure, high costs, and difficulties in ensuring traceability further complicate logistics. The seasonal nature of AFNs demands advanced warehousing, order processing, and storage solutions, which many small producers lack access to. Addressing these logistics challenges requires tailored approaches that differ significantly from the large-scale strategies used in traditional food systems.

To overcome the challenges of small-scale production, AFNs should explore collaborative logistics models ***to scale their operations***, such as for example *shared warehousing and transportation*. By pooling resources, AFNs can reduce costs and improve the efficiency of storage and transport.

The incorporation of ***digital technologies*** into AFNs enhances their operational efficiency, sustainability, and transparency. Digital tools such as route optimization software, supply chain management (SCM) systems, e-commerce platforms, digital tools for recyclable packaging, digital suitability labels, play a crucial role in improving logistics operations within SFSCs. These technologies help AFNs streamline transportation, reduce waste, and optimize resource use, contributing to a reduction in carbon emissions and operational costs.

Additionally, digital solutions promote ***transparency***, which is increasingly demanded by consumers who want to understand the journey of their food from "farm to fork". Therefore, embracing digital innovation is a key strategy for AFNs to remain competitive, meet consumer demands for transparency, and foster more sustainable, efficient, and resilient food supply chains.

While digitalization presents solutions to these logistics challenges, many AFNs struggle with adoption due to ***financial constraints***, limited technical expertise, and a lack of collaboration between stakeholders. Small-scale producers often find it difficult to invest in new technologies or utilize them effectively, particularly if they lack the necessary support and training. The need for scalable, flexible, and cost-effective digital solutions tailored to the specific needs of AFNs is critical for their long-term success.

Rural areas and short food supply chains require tailored innovation policies, but implementing these can be challenging for ***policymakers***. They can play a critical role in overcoming AFNs obstacles by creating programs that promote digital literacy, providing financial incentives, and ensuring that small producers have access to affordable digital tool (Fabregas et al. 2019). The synergy between policy frameworks like Farm to Fork (Commission 2024a) and the Common Agricultural Policy (Commission 2024b) and digital innovation is crucial to creating more resilient, transparent, and efficient short food supply chains. AFNs should actively seek government grants, subsidies, and partnerships with research institutions to help fund the adoption of digital solutions.

AFNs should leverage digital platforms to ***engage consumers*** and build stronger relationships. This can be achieved by digital sustainability labels to provide detailed information about product origins, ethical sourcing, and environmental impact and expanding e-commerce platforms to include features such as personalized recommendations and stories about local producers, enhancing consumer trust and satisfaction. Initiatives like community-supported agriculture (CSA) and local food hubs can be tested to enhance supply chain resilience and efficiency.

Future research should explore the practical implementation of emerging digital technologies such as blockchain and artificial intelligence in real-world AFNs, focusing on cost-effectiveness and scalability for small-scale producers. Studies examining the social impact of digital adoption, such as changes in consumer behaviour, trust-building, and producer-consumer relationships, could provide deeper insights into the human dimension of digital transformation. Additionally, living labs dedicated to food systems could play a pivotal role in co-creating and testing innovative solutions in collaboration with stakeholders, bridging the gap between theoretical advancements and practical applications. Comparative analyses of policy frameworks across regions can identify best practices for fostering digital literacy and innovation in short food supply chains. Finally, there is a need for long-term studies assessing the environmental and economic outcomes of integrating digital twins and collaborative logistics models into AFN operations, ensuring that sustain-ability goals are met alongside operational efficiency.

6 Conclusion

By embracing digital innovation and fostering collaboration, AFNs can overcome the unique logistics challenges they face. Tailored digital solutions, collaborative logistics models, and strategic policy support are essential to driving efficiency, sustainability, and transparency. With a focus on scalable digital technology and innovation, AFNs can strengthen their operations, enhance consumer trust, and build a more resilient and sustainable local food system for the future.

References

Akac A, Anagnostopoulou A, Nalmpantis D (2021) Digitalization in Freight Transport Services: Balkan Area BT - Advances in Mobility-as-a-Service Systems. In: Nathanail EG, Adamos G, Karakikes I (eds) Advances in Intelligent Systems and Computing. Springer International Publishing, Cham, pp 1056–1065

Albrecht T et al. (2024) Leveraging digital technologies in logistics 4.0: Insights on affordances from intralogistics processes. Inf Syst Front 26(2):755–774. https://doi.org/10.1007/s10796-023-10394-6

Arias-Aranda D, Bustinza OF, Barrales-Molina V (2011) Operations flexibility and outsourcing benefits: An empirical study in service firms. Serv Ind J 31(11):1849–1870. https://doi.org/10.1080/02642069.2010.503880

Ashtab S, Xing Y, Zheng C (2020) Exploring constituents of short food supply chains. In: Petrillo A et al (eds) Operations Management—Emerging Trend in the Digital Era. Available at: www.intechopen.com

Atalan-Helicke N, Abiral B (2021) Alternative food distribution networks, resilience, and urban food security in Turkey during the COVID-19 pandemic. J Agric Food Syst Community Dev 10(2):89–104. https://doi.org/10.5304/jafscd.2021.102.021

Baldwin R, Freeman R (2022) Risks and global supply chains: What we know and what we need to know. Annu Rev Econ 12(14):36. https://doi.org/10.1146/annurev-economics

Barbera F, Dagnes J (2016) Building alternatives from the Bottom-up: The case of alternative food networks. Agriculture and Agricultural Science Procedia 8, pp 324–331. https://doi.org/10.1016/j.aaspro.2016.02.027

Barbosa-Cánovas GV et al (2011) Food transportation issues and reducing carbon footprint. In: Joyce I. Boye, Yves Arcand (eds) Green Technologies in Food Production and Processing. Boston, MA: Springer. http://www.springer.com/series/5996

Bastard A, Chaillet A (2023) Digitalization from vine to wine: Successes and remaining challenges—A review. In BIO Web of Conferences. EDP Sciences. https://doi.org/10.1051/bioconf/20236801034

Bazzani C, Canavari M (2013) Alternative agri-food networks and short food supply chains: A review of the literature. Econ Agro-Aliment 15:1134. https://doi.org/10.3280/ECAG2013-002002

Behdani B, Fan Y, Bloemhof J (2019) Cool chain and temperature-controlled transport: An overview of concepts, challenges and technologies. In: Accorsi R, Manzini R (eds) Sustainable Food

Supply Chains. Academic Press, pp 167–183. Available at: https://research.wur.nl/en/publications/cool-chain-and-temperature-controlled-transport-an-overview-of-co. Accessed 14 Nov 2024

Berti G, Mulligan C (2016) Competitiveness of small farms and innovative food supply chains: The role of food hubs in creating sustainable regional and local food systems. Sustainability (Switzerland) 8(7). https://doi.org/10.3390/su8070616

Bio Austria (2024) Bio Austria organisation. https://www.bio-austria.at/bio-austria/. Accessed 19 Sep 2024

Borsellino V, Schimmenti E,.El Bilali H (2020) Agri-food markets towards sustainable patterns. Sustainability (Switzerland). MDPI. https://doi.org/10.3390/su12062193

Bos E, Owen L (2016) Virtual reconnection: The online spaces of alternative food networks in England. J Rural Stud 45:1–14. https://doi.org/10.1016/j.jrurstud.2016.02.016

Bosona T et al (2011) Integrated logistics network for the supply chain of locally produced food, Part I: Location and route optimization analyses. J Serv Sci Manag 04(02):174–183. https://doi.org/10.4236/jssm.2011.42021

Bosona T, Gebresenbet G (2013) Food traceability as an integral part of logistics management in food and agricultural supply chain. Food Control 32–48. https://doi.org/10.1016/j.foodcont.2013.02.004

Bräysy O et al (2009) An optimization approach for communal home meal delivery service: A case study. J Comput Appl Math 232(1):46–53. https://doi.org/10.1016/j.cam.2008.10.038

Brinkley C (2018) The smallworld of the alternative food network. Sustainability (Switzerland) 10(8). https://doi.org/10.3390/su10082921

Buhaug H, Von Uexkull N (2024) Annual review of environment and resources vicious circles: Violence, vulnerability, and climate change. https://doi.org/10.1146/annurev-environ-012220

Cagliano AC et al (2017) Analyzing the diffusion of eco-friendly vans for urban freight distribution. Int J Logist Manag 28(4):1218–1242. https://doi.org/10.1108/IJLM-05-2016-0123

Cao S, Johnson H, Tulloch A (2022) Exploring blockchain-based traceability for food supply chain sustainability: Towards a better way of sustainability communication with consumers, in procedia computer science. Elsevier B.V, pp 1437–1445. https://doi.org/10.1016/j.procs.2022.12.342

Cerrada-Serra P et al (2018) Exploring the contribution of alternative food networks to food security. A comparative analysis. Food Security 10:1371–1388. https://doi.org/10.1007/s12571-018-0860-x

Chi Ffoleau Y, Dourian T (2020) Sustainable food supply chains: Is shortening the answer? a literature review for a research and innovation agenda. Sustainability (Switzerland) 12(23):1–21. https://doi.org/10.3390/su12239831

Desalegn G, Tangl A, Fekete-Farkas M (2022) From Short-Term risk to Long-Term strategic challenges: Reviewing the consequences of geopolitics and COVID-19 on economic performance. Sustainability (Switzerland). MDPI. https://doi.org/10.3390/su142114455

Dominic C (2023) The importance of short food supply chain: Examining logistics strategies for competitive advantage and sustainability, in agricultural value chains. https://doi.org/10.5772/intechopen.110429

Edwards F (2016) Alternative food networks. In: Thompson P, Kaplan D (eds) Encyclopedia of Food and Agricultural Ethics. Dordrecht: Springer

Esteso A et al (2022) Collaborative optimization tool for sustainable planning of an agricultural supply chain preserving farmers' Independence BT—Collaborative networks in digitalization and society 5.0'. In: Camarinha-Matos LM et al (eds) Collaborative Networks in Digitalization and Society 5.0. Cham: Springer International Publishing, pp 519–532

European Commission (2024a) Farm to Fork strategy. https://food.ec.europa.eu/horizontal-topics/farm-fork-strategy_en Accessed 19 Sep 2024

European Commission (2024b) The common agricultural policy at a glance. Available at: https://agriculture.ec.europa.eu/common-agricultural-policy/cap-overview/cap-glance_en. Accessed 19 Sep 2024

European Parliament (2016) Short food supply chains and local food systems in the EU. https://www.europarl.europa.eu/ Accessed 20 Sep 2024

Fabregas R, Kremer M, Schilbach F (2019) Realizing the potential of digital development: The case of agricultural advice. Science 366(6471). https://doi.org/10.1126/science.aay30

Fares N et al (2024) A hybrid framework for fleet management with quality concerns: A case for the food industry. Int J Qual Reliab Manag 41(9):2351–2372. https://doi.org/10.1108/IJQRM-08-2022-0241

Fikar C, Leithner M (2021) A decision support system to facilitate collaborative supply of food cooperatives. Prod Plan Control 32(14):1179–1190. https://doi.org/10.1080/09537287.2020.1796135

Forno F et al (2024) From grassroots to platforms: How Digitalisation reconfigures learning and engagement with food. In: Dulsrud A, Forno F (eds) Digital food provisioning in times of multiple crises. Palgrave Macmillan Cham. https://doi.org/10.1007/978-3-031-46323-5_7

Furtado P, Montréal P, Frayret J-M (2014) Impact of resource sharing of freight transportation.1st International Physical internet conference. https://www.researchgate.net/publication/262901180

García-Díez J et al (2021) Determining food stability to achieve food security. Sustainability (Switzerland). MDPI. https://doi.org/10.3390/su13137222

Görçün ÖF et al (2024) Assessing and selecting sustainable refrigerated road vehicles in food logistics using a novel multi-criteria group decision-making model. Inf Sci 661. https://doi.org/10.1016/j.ins.2024.120161

Gori F, Castellini A (2023) Alternative food networks and short food supply chains: A systematic literature review based on a case study approach. Sustaiability 15(10). https://doi org/10.3390/su15108140

Goszczynski W et al (2019) Between imitation and embeddedness: Three types of polish alternative food networks. Sustainability (Switzerland) 11(24). https://doi.org/10.3390/su11247059

Govindan K et al (2014) Two-echelon multiple-vehicle location-routing problem with time windows for optimization of sustainable supply chain network of perishable food. Int J Prod Econ 152:9–28. https://doi.org/10.1016/j.ijpe.2013.12.028

Hausman A, Johnston WJ (2014) The role of innovation in driving the economy: Lessons from the global financial crisis. J Bus Res 67(1):2720–2726. https://doi.org/10.1016/j.jbusres.2013.03.021

Inkinen T, Hämäläinen E (2020) Reviewing truck logistics: Solutions for achieving low emission road freight transport. Sustainability (Switzerland) 12(17). https://doi.org/10.3390/SU12176714

Jagtap S et al (2021) Food logistics 4.0: Opportunities and challenges. Logistics 5(1). https://doi.org/10.3390/logistics5010002

Jia F et al (2024) Promoting resilient and sustainable food systems: A systematic literature review on short food supply chains. J Clean Prod. Elsevier Ltd. https://doi.org/10.1016/j.jclepro.2023.140364

Kapała AM (2022) Legal instruments to support short food supply chains and local food systems in France. Laws 11(2). https://doi.org/10.3390/laws11020021

Kneafsey M et al (2013) Short food supply chains and local food systems in the EU. A state of play of their Socio-Economic characteristics

Kondo C, Suzuki A (2022) The pitfalls within alternative food networks: A comparison between Japan's wholesale market system and alternative market distribution challenges. In: Mars MM, Schau HJ (eds) How Alternative is Alternative? The Role of Entrepreneurial Development, Form, and Function in the Emergence of Alternative Marketscapes. Leeds: Emerald Publishing Limited, pp 123–147. https://doi.org/10.1108/S1048-473620220000029006

Laginová L, Hrivnák M, Jarábková J (2023) Organizational models of alternative food networks within the Rural–Urban interface. Adm Sci 13(9). https://doi.org/10.3390/admsci13090193

Lezoche M et al (2020) Agri-food 4.0: A survey of the supply chains and technologies for the future agriculture. Comput Ind. Elsevier B.V. https://doi.org/10.1016/j.compind.2020.103187

Liu TC, Wu YC, Chau CF (2023) An overview of carbon emission mitigation in the food industry: Efforts, challenges, and opportunities, Processes. Multidisciplinary Digital Publishing Institute (MDPI). https://doi.org/10.3390/pr11071993

Mahendran K et al (2024) A systematic review on the delivery optimization in food delivery industry. In: Proceedings of Conference on Transdisciplinary Research in Engineering

Martindale L (2021) I will know it when I taste it: Trust, food materialities and social media in Chinese alternative food networks. Agric Hum Values 38(2):365–380. https://doi.org/10.1007/s10460-020-10155-0

Mcclements DJ et al (2020) Building a resilient, sustainable, and healthier food supply through innovation and technology. Annu Rev Food Sci Technol 12:56. https://doi.org/10.1146/annurev-food-092220

Mecalux (2023) Logistics processes: Key points for optimising your supply chain. https://www.mecalux.com/blog/logistics-processes. Accessed 19 Sep 2024

Michel-Villarreal R et al (2019) Sustainability in alternative food networks: A systematic literature review. Sustainability (Switzerland). MDPI. https://doi.org/10.3390/su11030859

Michel-Villarreal R et al (2021) Resilience and digitalization in short food supply chains: A case study approach. Sustainability (Switzerland) 13(11). https://doi.org/10.3390/su13115913

Mittal A, Krejci CC, Craven TJ (2018) Logistics best practices for regional food systems: A review. Sustainability (Switzerland). MDPI. https://doi.org/10.3390/su10010168

Miwa T, Bell MGH (2017) Efficiency of routing and scheduling system for small and medium size enterprises utilizing vehicle location data. J Intell Transp Syst 21(3):239–250. https://doi.org/10.1080/15472450.2017.1291350

Mondino P, Gonzalez-Andujar JL (2019) Evaluation of a decision support system for crop protection in apple orchards. Comput Ind 107:99–103. https://doi.org/10.1016/j.compind.2019.02.005

Ibiyemi MO, Olutimehin DO (2024) Utilizing predictive analytics to enhance supply chain efficiency and reduce operational costs. Int J Eng Res Updat 7(1):01–021. https://doi.org/10.53430/ijeru.2024.7.1.0029

Nordmark I et al (2012) Integrated logistics network for the supply chain of locally produced food, Part II: Assessment of E-trade, economic benefit and environmental impact. J Serv Sci Manag 05(03):249–262. https://doi.org/10.4236/jssm.2012.53030

Opitz I et al (2017) Effects of consumer-producer interactions in alternative food networks on consumers learning about food and agriculture. Moravian Geographical Reports 25(3):181–191. https://doi.org/10.1515/mgr-2017-0016.

Oriekhoe OI et al (2024) Review of technological advancements in food supply chain management: A comparative study between the U.S. and Africa. Int J Manag Entrep Res 6(1):132–149. https://doi.org/10.51594/ijmer.v6i1.713

Parrag V, Fricz Szegedyné Á, Sebők A (2022) Application of digital solutions to improve the operation of short food supply chains. Int J Food Stud 11(2)

Petruzzelli M et al (2023) The role of short food supply chains in advancing the global agenda for sustainable food systems transitions. Cities 141

Pinna S (2016) Alternative food networks, agro-biodiversity and landscape protection: Lessons from two rural parks. Reg Sci 3(1):455–462. https://doi.org/10.1080/21681376.2016.1244488

Poças Ribeiro A et al (2021) Organising Alternative Food Networks (AFNs): Challenges and facilitating conditions of different AFN types in three EU countries. Sociol Rural 61(2):491–517. https://doi.org/10.1111/soru.12331

Reckinger R (2022) Values-based territorial food networks qualifying sustainable and ethical transitions of alternative food networks. 12(3):78–109. https://doi.org/10.3167/reco.2022.120305

Rogers ZS et al (2020) A Sharing-Enabled portfolio approach to distribution. Transp J 59(2):99–128. https://doi.org/10.5325/transportationj.59.2.0099

Singh S et al (2024) Inventory management using big data and internet of things. https://engrxiv.org/preprint/view/3707. Accessed 14 Nov 2024

Skalkos D (2024) Scientific advancements for an innovative agri-food supply chain towards the 2030 sustainable development goals III. Sustainability (Switzerland) . Multidisciplinary Digital Publishing Institute (MDPI). https://doi.org/10.3390/su16135693

Stein AJ, Santini F (2022) The sustainability of "local" food: A review for policy-makers. Rev Agric Food Environ Stud 103(1):77–89. https://doi.org/10.1007/s41130-021-00148-w

Szabo Ľ, Richnák P, Gubová K (2021) New dimension of logistics innovations development in agricultural enterprises in Slovakia. Agricultural Economics (Czech Republic) 67(4):136–143. https://doi.org/10.17221/444/2020-AGRICECON

Todorovic V et al (2018) Solutions for more sustainable distribution in the short food supply chains. Sustainability (Switzerland) 10(10). https://doi.org/10.3390/su10103481

Tregear A (2011) Progressing knowledge in alternative and local food networks: Critical reflections and a research agenda. J Rural Stud 27(4):419–430

Vanderroost M et al (2017) The digitization of a food package's life cycle: Existing and emerging computer systems in the logistics and post-logistics phase. Computers in Industry. Elsevier B.V, pp 15–30. https://doi.org/10.1016/j.compind.2017.01.004

Viciunaite V (2023) Alternative food networks on digital platforms: Consumer preferences for platform versus local food attributes. J Rural Stud 100. https://doi.org/10.1016/j.jrurstud.2023.103017

Woolley E, Jellil A, Simeone A (2020) Wasting less food: Smart mass customisation of food provision. In Procedia CIRP. Elsevier B.V, pp 189–194. https://doi.org/10.1016/j.procir.2021.01.073

Xenou E et al (2021) The Rise of the On-Demand Warehousing: Is the Greek Market Ready for This Change? BT - Advances in Mobility-as-a-Service Systems. In: Nathanail EG, Adamos G, Karakikes I (eds) Advances in Mobility-as-a-Service Systems. Springer International Publishing, Cham, pp 835–844

El Yaagoubi A et al (2024) Towards Sustainable urban logistics: An optimization-simulation approach for electric vehicles routing in short food supply chains BT—Service Oriented, Holonic and Multi-Agent Manufacturing Systems for Industry of the Future. In: Borangiu T et al (eds) Cham: Springer Nature Switzerland, pp 212–224

Zhong R, Xu X, Wang L (2017) Food supply chain management: Systems, implementations, and future research. Ind Manag Data Syst 117(9):2085–2114. https://doi.org/10.1108/IMDS-09-2016-0391

Use of Artificial Intelligence in Interdisciplinary Risk Management in Road Transport

Agnieszka A. Tubis and Kristina Čižiūnienė

Abstract An analysis of publications on the use of AI in industry shows that one of its most important applications is in improving decision-making processes. However, it is worth noting that AI tools can act here in a dual role: (a) as a tool that supports human decision-making (helping humans) or (b) as a system that makes decisions on its own (replacing humans). Risk assessment and management are two analytical areas in which there is growing interest in the application of AI. Currently, most of the literature in this area is concerned with implementing AI tools and machine counting in risk management processes in the financial sector and risk estimation in disaster risk management. In contrast, the number of publications describing the current possibilities of applying AI in operational risk management in the road transport sector is limited. Therefore, the purpose of the publication is to explore the areas of research on the application of artificial intelligence in transport systems, particularly in road transport, and on this basis, to identify areas of application of AI tools in the road transport risk management process at strategic and operational levels. The results presented show the potential for implementing AI in risk analysis and proceedings aimed at minimising risks at both the system (engineering) and operational (process) levels.

Keywords Risk assessment · Risk mitigation · Process approach · Decision support

A. A. Tubis (✉)
Faculty of Mechanical Engineering, Wroclaw University of Science and Technology, Wroclaw, Poland
e-mail: agnieszka.tubis@pwr.edu.pl

K. Čižiūnienė
Department of Logistics and Transport Management, Vilnius Gediminas Technical University, Vilnius, Lithuania
e-mail: kristina.ciziuniene@vilniustech.lt

A. Kolinski and M. Adamczak (eds.), *Digitalisation of the Greening Supply Chain*, EcoProduction, https://doi.org/10.1007/978-3-031-88918-9_2

1 Introduction

A wide variety of definitions of artificial intelligence can be found in the literature. However, most of them refer to the core competencies of non-human intelligence, which is used to perform specific tasks according to an accepted algorithm (Dwivedi et al. 2021). In the research presented in (Russell and Norvig 2016), the authors characterised AI as systems that mimic cognitive functions defined as human attributes, such as learning, speech, and problem-solving. A step further was taken by Kaplan and Haenlein (2019), who described AI in the context of its ability to learn and independently interpret external data to achieve specific outcomes through flexible adaptation. Dwivedi et al. (2021) note that the common denominator of currently formulated definitions of AI is the ability of machines to perform specific tasks that humans currently perform in both the workplace and society. Additionally, these authors note that AI's ability to transcend human limitations on computational, intellectual and creative capabilities opens up new areas for its applications in education, marketing, healthcare, finance and manufacturing. Their published research findings indicate that currently (Dwivedi et al. 2021):

- The use of AI in the form of chatbots in customer service applications is growing.
- Technology giants (e.g. Walmart and Amazon) are testing AI to forecast demand and support supply chain flows.
- Walmart shops are testing the use of AI to analyse shopping baskets to trigger a response when customers select the last item and then be able to track the shop's ability to replenish product inventory quickly.

Today's organisations increasingly use AI solutions in their decision-making processes, opportunities are being sought to automate the work effectively, and some predict that artificial intelligence will replace most of the day-to-day duties of some white-collar workers (Du et al. 2021). At the same time, companies are increasingly integrating AI into selected processes to improve their productivity and the organisation's global efficiency (Hossain et al. 2022; Li et al. 2022; Liu and Li 2022). However, it is worth noting that the current level of development of AI technology does not justify considering it as a complete replacement for human labour in the enterprise, which determines that the so-called artificial general intelligence (AGI) is still a vision for the future (Dwivedi et al. 2021). Currently, many organisations see AI as an efficient and cost-effective way to automate certain activities and realise human-AI collaboration (HAC), especially in knowledge-intensive activities (Zheng et al. 2023). However, as emphasised by many researchers (Makarius et al. 2020; Østerlund et al. 2021; Seeber et al. 2020), a critical prerequisite for achieving the intended improvements in efficiency and productivity of integrated AI is excellent collaboration between AI and organisational employees. Therefore, it is necessary to create a new human-machine symbiosis and seek ways to collaborate to complement each other in organisational decision-making to enhance and improve their capabilities (Jarrahi 2018). The basis in this case is to leverage the different strengths of humans and AI in mutual collaboration (Herrmann and Pfeiffer 2023).

Many researchers describe value creation based on the application of AI in organisations. Noteworthy in this case is the research of Liu et al. (2023) in applying AI in the power industry. The results they describe indicate four potential application areas for AI, which may also provide guidance for the potential implementation of AI tools in other sectors (Liu et al. 2023):

- Prediction and optimisation of energy demand. The amount of energy demand is influenced by many factors that result in a high randomness and variability of loads. This makes its prediction a very complex process. In the context of big data, deep learning algorithms can return results that guarantee greater accuracy and robustness in predicting future loads.
- Smart grid management. Artificial intelligence can analyse data and learn effectively from it to quickly identify anomalies in the grid. On this basis, it is possible to use machine learning methods to provide early warning of, for example, the occurrence of energy theft.
- Energy trading and market forecasting. Deep learning can be used to integrate demand data from different power grids. In this way, improved prediction of emerging loads can be achieved.
- Customer service and personalised recommendations. The use of AI enables the implementation of an electricity price recovery strategy based on user reputation. This requires improving the ability of companies to control energy recovery risks based on complex risk assessment models.

An analysis of publications on the application of AI in industry shows that one of its most important applications is the improvement of decision-making processes. It is worth noting, however, that AI tools can play a dual role here: (a) as a tool that supports human decision-making (helping humans) or (b) as a system that makes decisions on its own (replacing humans) (Edwards et al. 2000). The continuous development of technology now enables researchers to create increasingly sophisticated machines, favouring AI's implementation in increasingly complex tasks, even those requiring cognitive abilities such as making tacit judgements, sensing emotions and directing processes (Mahroof 2019). At the same time, the observed steady development of AI tools has changed how information is generated and used for decision-making (Vaio et al. 2020). Some researchers emphasise that the application of AI in industry has changed how many companies do business (Schneider and Leyer 2019). Indeed, its implementation influences companies' business and management practices to offer more competitive and sustainable products and services (Govindan et al. 2019; Kuo and Smith 2018; Ding et al. 2019; Garbuio and Lin 2019; Tuan et al. 2019). Nevertheless, as they point out (Ellefsen et al. 2019), AI is a major challenge for manufacturing companies and the transport and logistics industry. However, These authors argue that AI should be increased and extended to solve practical problems. Therefore, mathematical algorithms that use large numbers of data, connections, and interactions are increasingly becoming part of standard organisational management and simplifying the decision-making process for managers (Schneider and Leyer 2019).

Risk assessment and management are two analytical areas in which there is growing interest in the application of AI. Currently, most publications in this area are concerned with implementing AI tools and machine counting in risk management processes in the financial sector (Aziz and Dowling 2018; Fritz-Morgenthal et al. 2022) and risk estimation in disaster risk management (e.g. (Gevaert et al. 2021)). Artificial intelligence can support institutions at various stages of risk management through detailed analysis of available source data (Nagy et al. 2022). It can be used to comprehensively identify undesirable events, estimating their probability of occurrence and impact. However, AI can also support selecting an appropriate risk mitigation strategy or finding instruments that effectively facilitate the transfer or trading of risks (Liu et al. 2025; Gursel et al. 2025). Aziz and Dowling (Aziz and Dowling 2018) point to the extended use of AI in risk management, including the analysis of extensive sets of documents, the execution of repetitive processes, and the detection of money laundering, all of which require the analysis of large data sets.

A preliminary literature survey conducted by the authors indicates that the number of publications describing the current applicability of AI in operational risk management in the road transport sector is limited. Therefore, the purpose of the publication is to explore the areas of research on the application of artificial intelligence in transport systems, particularly in road transport, and on this basis, identify areas of application of AI tools in road transport risk management at the strategic and operational levels. The main contribution of the presented research results is:

- A comprehensive literature review of the application of artificial intelligence in selected transportation systems.
- Identification of areas of application of artificial intelligence tools in road transport and identification of the main challenges of their implementation based on literature studies.
- Identify the potential for applying artificial intelligence in risk management in road transportation based on the engineering-process approach.

The structure of the publication includes Sect. 2, which characterizes risk management systems in road transportation in engineering and process approach. Section 3 presents the methodology of the conducted research, which consists of 4 stages. Subsequently, Sect. 4 presents the results from the completed research work and answers the research questions posed. Section 5 formulates the conclusions and provides directions for further research development in this area.

2 Risk Management in Road Transport in Engineering and Process Approach

Different approaches to risk management can be found in the literature, the diversity of which is mainly due to the subject of analysis, the defined context of risk assessment and the objectives to be served by the analytical process. Concerning transport systems, two approaches—engineering and managerial (derived from the concept of Enterprise Risk Management)—become critical.

In the engineering approach, the concept of risk is very strongly linked to the concept of safety management in relation to the operation of technical systems. For this reason, the risk is viewed exclusively from a negative aspect (in line with the so-called German school (Tubis 2018)). Therefore, its occurrence is associated with undesirable consequences, such as loss or damage. The distinction between these two concepts is shown in Table 1.

The second critical approach to risk management is the managerial approach, which was disseminated by the COSO organisation as the Enterprise Risk Management (Institute of risk management 2018) concept and became the basis for the created ISO 31000 and ISO 31010 (ISO 31000 2018; ISO 31010 2019). According to this standard, risk is defined as 'the impact of uncertainty on objectives'. However, it should be emphasised that this impact can be both negative and positive. This approach is in line with the so-called American school (Tubis 2018), according to which the impact of risk can be loss or gain.

The detailed study of risk issues in various transport systems presented in (Tubis 2018) shows that the analyses and results described are exclusively oriented towards an engineering approach and primarily refer to safety-related aspects. This is even confirmed by the risk definitions formulated, which refer to the occurrence of a loss and equate risk with an accident. Examples of such definitions for road transport are given below:

- Risk is the probability of an accident (Hauer 1982).
- Risk is the probability and consequences of an accident (Haight 1986).

Although the above definitions date from the late 20th century, this understanding of risk in road transport is still valid in research and should be considered the dominant one in the world literature. It should also be noted that in this approach, the focus of the research is the protection of the health and life of road users, and the analyses carried out relate to the functioning of the transport system. Therefore, this level of risk assessment and management is strategic.

For risk assessment in road transport at the strategic level, Smalko, Nowakowski and Tubis (2020) proposed an anthropo-techno-sphere approach. In this approach, the identification of adverse events carried out covers three areas:

Table 1 Distinction between loss and damage as consequences of adverse events—based on: (Smalko et al. 2020)

	Damage	Loss
Assessment method	Qualitative assessment	Quantitative assessment
Means of expressing consequences	Damage to human health, destruction of material objects or degradation of the environment	Loss of possession expressed in material or financial terms

- Humans—primarily the vehicle's driver and other road users may also be the subject of analysis.
- Vehicle—primarily in terms of reliability and damage occurring.
- Environment—analysed from two aspects: (a) the impact of road conditions and weather phenomena and (b) the transport infrastructure.

A review of the literature indicates that driving errors are considered to be the leading cause of accidents. The type of errors and their frequency depend primarily on the drivers' experience but also the driver's level of fatigue and risk perception. The most common driver error is not adjusting the speed to the road conditions. For this reason, many studies have looked at average speed and speed variance in terms of their relevance and occurrence (Hussain et al. 2019; Doecke et al. 2018; Aarts and Schagen 2006). The second significant risk is driver fatigue. The main effect of driver fatigue is a gradual loss of attention and lack of concentration, which results in a deterioration of driving quality (Brown 1997). Studies of the causes of fatigue (including lack of required sleep time) and its effects are presented in (Williamson and Feyer 2000; Sagberg 1999; Oron-Gilad and Shinar 2000), among others.

Vehicle risk assessment is usually part of technical reliability analyses or the determination of fleet maintenance strategies. Examples of such publications are (Niewczas et al. 2023). Research in this context is also concerned with the early identification of signals of potential damage, such as (Kowalik and Szpyrka 2019). It should also be noted that these topics are also addressed in research on improving vehicle safety (Budd and Newstead 2021).

The influence of the environment on adverse events is primarily assessed in terms of the prevailing road conditions. Weather factors are an important determinant of these conditions. The most frequently analysed weather phenomenon is precipitation, which significantly influences unsafe transport behaviour (Ettema et al. 2017; Satterthwaite 1976). Another weather parameter analysed in studies related to road transport risks is the occurrence of fog (Al-Ghamdi 2007; Abdel-Aty et al. 2011) or strong winds (Levine et al. 1995; Baker and Reynolds 1992; Brijs et al. 2008). The influence of the environment on the emergence of hazardous situations is also analysed in the context of natural disasters such as earthquakes (Dong et al. 2014; Dong et al. 2013) and landslide floods (Clarke and Obrien 2016).

The second element of the environment is the road infrastructure in use. However, as studies indicate, the poor quality of the road and its surroundings is rarely indicated in statistics as a direct cause of road accidents (Tubis 2018). For this reason, most research on road infrastructure risks focuses on identifying the risks associated with tunnels (Amundsen and Ranes 2000; Vuilleumier et al. 2002) and bridges (Wang et al. 2008; Aziz and Kodur 2013).

When assessing risk in road transport, it is also important to consider the process approach recommended in ISO 31000 (2018). For this reason, Tubis, in her research described in (Tubis 2018), proposed an operational approach to risk assessment in road transport processes. This risk was defined as the occurrence of undesirable events which disrupt the correct course of the process or limit the achievement of the adopted operational objective (Tubis 2018). When assessing risk for transport systems, it is critical to consider the perspective of business process management. This perspective means that the criteria for assessing the degree to which objectives are met are not just the health and lives of process participants but the time, cost and quality of operations performed.

For this reason, operational hazards in the road transport process are defined according to the business continuity management standard as a negative consequence resulting from an incident that affects the organisation's objectives (British Standard and Institution 2007). Therefore, the starting point for identifying potential adverse incidents is the adopted objectives, which are the assumed results of the process. In the case of the road transport process, this is the delivery of a transport service with a defined cost of delivery and defined logistical and quality parameters. Any event negatively affecting the achievement of such a set objective should be subject to a risk analysis (Smalko et al. 2020).

3 Methodology

The research conducted on the application of artificial intelligence in the field of interdisciplinary risk management in road transport is primarily theoretical, and it aims to identify the potential associated with the technological development of AI-based tools that can support road transport processes. For this reason, the research carried out is based on various sources of knowledge and is of an initial research nature, which the authors will develop as part of their ongoing research work. The analyses and inferences prepared aimed to answer the following research questions:

- Q1: What are the current areas of application of AI in transport systems reported in the literature?
- Q2: Which of the applied AI solutions can be used to manage risk at a strategic level?
- Q3: In which areas can AI tools be applied to manage risk at an operational level?
- Q4: How can applying AI tools affect the operational risks present in road freight handling processes?
- To answer the questions formulated in this way, a 4-step research procedure was applied, shown in Fig. 1, together with an indication of the sources of knowledge used in the research carried out.

The literature review was prepared using the narrative review method, which does not have a structured document search strategy but focuses on the research question (the research question posed). The search for papers describing the use of AI in transport was conducted using the Scopus and Web of Science databases. The following keywords were used to identify publications: 'AI', 'artificial intelligence', 'risk', and 'transport'. Then, from the set of papers, those articles describing results concerning AI's application in road transport were highlighted. Based on the set of papers defined in this way, those articles that described results concerning the use of AI in risk management were selected. Based on the analysis of the benefits and

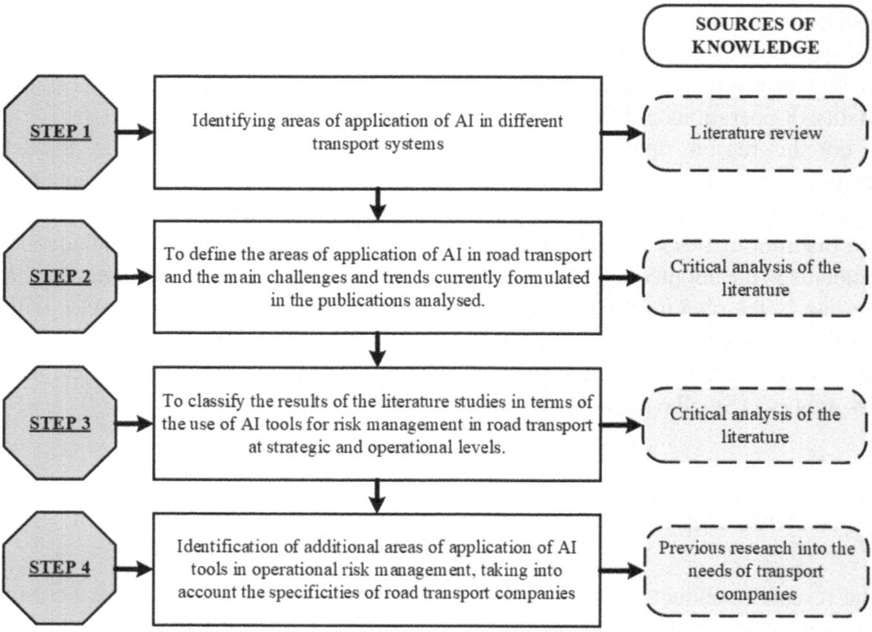

Fig. 1 Flow chart of the research, own study

limitations of AI identified in the literature, areas of application of AI in operational risk management in road transport were proposed. The research carried out following the established procedure provided knowledge to answer the formulated questions. The results obtained are presented in Sect. 4.

4 Results

4.1 The Application of AI in Various Transport Sectors and in Transportation Systems

The application of artificial intelligence (AI) in various transportation sectors is revolutionizing how people and goods move around the world. Advances in AI are enhancing efficiency, safety, and sustainability in road, air, maritime, and rail transportation.

Artificial intelligence (AI) is fundamentally reshaping transportation systems around the world. Integrating AI into transportation infrastructures is driving innovation across various modes of travel, including road, rail, air, and maritime transport. This transformation enhances operational efficiency, improves safety, reduces environmental impact, and creates new possibilities for intelligent, responsive transportation networks. The following sections explore the multifaceted applications of AI in transportation systems.

The literature review results are presented from two perspectives—in transport sectors and transport systems. When comparing the application of AI in various transport sectors and transport systems, it is possible to single out several essential criteria. For this to be possible, however, it is necessary to clarify the relevance of the two perspectives as a first step:

- The transport sector is a category of companies that provide services for the transportation of people or goods, as well as transportation infrastructure. Technically, transportation is a subgroup of the industry sector under the Global Industry Classification Standard (GICS) (https://www.investopedia.com/terms/t/transportation_sector.asp).
- A transportation system can be defined as the combination of elements and their interactions, which produce the demand for travel within a given area and the supply of transportation services to satisfy this demand (Cascetta 2001).

Based on this division, a summary of results has been prepared in the form of Table 2.

Summarizing the information presented in Table 2, it can be said that when examining both the various transport sectors and transport systems, there is a tendency to describe the application of AI with similar elements, with minimal differences in quantity or content.

Despite these minimal differences, it is important to assess the impact and challenges, as well as various considerations about the future prospects of AI applications.

The impact of AI on various transportation sectors is significant, with benefits including increased efficiency, enhanced safety, and cost reduction. However, there are also challenges (of transportation sectors and transportation systems) that need to be addressed for widespread and effective adoption of these technologies (Table 3).

In conclusion, it can be said that AI is increasingly being used both in various transport sectors and in transport systems.

After evaluating the general criteria, it is important to analyze the possibilities and challenges of applying AI in road transportation systems.

4.2 *The Application of AI in Road Transportation Systems*

Artificial intelligence (AI) is profoundly transforming road transportation systems, creating more efficient, safer, and sustainable modes of travel. Integrating AI into various aspects of road transportation, including autonomous vehicles, traffic management, smart infrastructure, and predictive maintenance, is driving significant advancements. Based on the literature research, key areas of its application in road transport relevant to risk management were identified:

Table 2 A comparison of the application of AI in different transport sectors and transport systems, own elaboration based on (Ahangar et al. 2021; Garikapati and Shetiya 2024; Pipicelli et al. 2024; Jain et al. 2021; Cheng et al. 2022; Jun et al. 2023; Dauptain et al. 2022; Gu et al. 2023; Manzoor et al. 2024; Guo et al. 2023; Gheluwe et al. 2020; Qiao et al. 2022; Ravish and Swamy 2021; Rojek et al. 2023; Kurek and Macioszek 2021; Saleem et al. 2020; Ćelić et al. 2024; Stopka et al. 2020; Stoilova 2025; Li and Zhu 2022; Li et al. 2022; Chen et al. 2022; Song et al. 2023; Deng et al. 2020; Cuppi et al. 2021; Bretas et al. 2023; Zhang and Zhang 2023; Luo et al. 2022; Smoczyński et al. 2021; Garmabaki et al. 2021; Bureika et al. 2022; Aydoğan and Cetek 2023; Hildemann and Verstegen 2023; Szabo et al. 2020; Kumar et al. 2022; Eskenazi et al. 2023; Lee et al. 2023; Stanton et al. 2023; Pater et al. 2022; Heim et al. 2020; Lee and Mitici 2022; Carr et al. 2020; Karamitsos et al. 2023; Naji et al. 2020; Janssen et al. 2020; Wang et al. 2022, 2023; Zhang and Zhang 2023; Hwang and Youn 2022; He et al. 2022; Munim et al. 2022; Dai et al. 2024; Mollaoglu et al. 2023; Li et al. 2022; Öztürk and Başar 2022; Scalia et al. 2023; Zhen et al. 2020; Makridis et al. 2020; Halvorsen-Weare and Nonås 2023; Angelopoulos et al. 2024; Liu et al. 2022)

	AI in different transport sectors	AI in transportation systems
Road transport	*Autonomous vehicle* (Ahangar et al. 2021; Garikapati and Shetiya 2024; Pipicelli et al. 2024; Jain et al. 2021; Cheng et al. 2022; Jun et al. 2023; Dauptain et al. 2022)	
	Autonomous vehicles are perhaps the most visible and promising application of AI in land transportation. These systems use deep learning algorithms to process vast amounts of data from sensors, cameras, and radars, enabling vehicles to recognize and react to their surroundings in real time	Autonomous vehicles (AVs) are at the forefront of AI applications in road transport. These vehicles can navigate complex environments, detect obstacles, and make real-time decisions, significantly reducing the likelihood of human error-related accidents
	Traffic management (Gu et al. 2023; Manzoor et al. 2024; Guo et al. 2023; Gheluwe et al. 2020; Qiao et al. 2022; Ravish and Swamy 2021)	
	AI is used to optimize traffic management in cities. Intelligent traffic systems collect real-time data from cameras, sensors, and GPS devices to analyze traffic patterns and adjust traffic lights accordingly. This helps reduce congestion, improve traffic flow, and lower carbon emissions	AI-powered traffic management systems are revolutionizing how cities handle congestion. These systems use data from traffic cameras, sensors, and GPS devices to analyze real-time traffic conditions. AI algorithms then optimize traffic light timings, predict congestion patterns, and suggest alternative routes to drivers
	Predictive maintenance (Rojek et al. 2023)	
	Predictive maintenance systems use AI to predict when vehicle and infrastructure failures are likely to occur. By analyzing data from sensors that monitor the performance of critical components, AI algorithms can anticipate failures and schedule maintenance before costly breakdowns occur. This increases the reliability and safety of land transportation	Predictive maintenance in road transport involves using AI to monitor vehicle conditions and predict when maintenance is needed. By analyzing data from various vehicle sensors, AI can identify potential issues before they become critical, allowing for timely repairs. This reduces downtime, extends vehicle lifespans, and enhances safety
	Smart parking (Kurek and Macioszek 2021; Saleem et al. 2020; Ćelić et al. 2024)	

(continued)

Table 2 (continued)

	AI in different transport sectors	AI in transportation systems
	–	AI is also enhancing parking solutions in urban areas. Smart parking systems use AI to predict parking space availability and guide drivers to the nearest open spots. These systems rely on real-time data from cameras and sensors, helping reduce the time spent searching for parking and minimizing congestion and emissions in city centres
Rail Transport	*Autonomous train and control* (Stopka et al. 2020; Stoilova 2025; Li and Zhu 2022; Li et al. 2022; Chen et al. 2022; Song et al. 2023; Deng et al. 2020)	
	AI is used for autonomous train control. Autonomous train technology is being implemented in various parts of the world. These systems use AI to control the speed, acceleration, and braking of trains, optimizing energy consumption and enhancing safety	AI enables the development of autonomous trains, which can operate with minimal human intervention. These trains use AI to control speed, braking, and acceleration, optimizing energy consumption and improving punctuality
	Rail traffic management (Cuppi et al. 2021; Bretas et al. 2023; Zhang and Zhang 2023; Luo et al. 2022)	
	Rail Traffic Management: AI is also used to manage rail traffic more efficiently. AI-based systems can analyze real-time data from railway operations and adjust train schedules to minimize delays and improve punctuality. This is especially useful in densely populated rail networks, where traffic optimization is crucial to maintaining service	Rail Traffic Management: AI-driven rail traffic management systems analyze data from train operations to optimize scheduling and routing. These systems can adjust train timings to minimize delays and ensure smooth operations across the network. This is particularly beneficial in densely populated areas where efficient rail traffic management is critical for maintaining reliable service
	Infrastructure maintenance (Smoczyński et al. 2021; Garmabaki et al. 2021; Bureika et al. 2022)	

(continued)

Table 2 (continued)

	AI in different transport sectors	AI in transportation systems
	Predictive maintenance plays an important role in the rail sector as well. Sensors installed on tracks and trains collect data on the condition of infrastructure. This data is analyzed to identify potential issues before they become serious failures, allowing for more efficient maintenance and reducing downtime	Predictive maintenance is crucial for rail systems as well. AI analyzes data from sensors placed on tracks and trains to detect anomalies and predict infrastructure failures. This allows for proactive maintenance, reducing the risk of disruptions and ensuring the safety and reliability of rail services
Air transport	*Flight route optimization* (Aydoğan and Cetek 2023; Hildemann and Verstegen 2023; Szabo et al. 2020; Kumar et al. 2022; Eskenazi et al. 2023)	
	AI is used to optimize flight routes, helping to reduce fuel consumption and carbon emissions. Algorithms analyze weather data, air traffic patterns, and flight conditions to determine the most efficient routes	AI is transforming air transportation by optimizing flight routes. By analyzing weather patterns, air traffic data, and aircraft performance, AI algorithms can suggest the most efficient routes, reducing fuel consumption and emissions. This not only lowers operational costs for airlines but also mitigates the environmental impact of air travel
	Aircraft predictive maintenance (Lee et al. 2023; Stanton et al. 2023; Pater et al. 2022; Heim et al. 2020; Lee and Mitici 2022)	
	Similar to the land sector, AI is applied in the predictive maintenance of aircraft. Sensors installed on aeroplanes collect real-time data on engine performance and other critical systems. This data is analyzed using machine learning algorithms to foresee potential failures and plan maintenance proactively, reducing downtime and improving safety	AI is also applied in predictive maintenance for aircraft. Sensors installed on aeroplanes collect real-time data on engine performance and other critical systems. Machine learning algorithms analyze this data to predict potential failures, enabling airlines to conduct maintenance before issues become serious. This improves safety, reduces downtime, and lowers maintenance costs
	Airport security and operations (Carr et al. 2020; Karamitsos et al. 2023; Naji et al. 2020; Janssen et al. 2020)	

(continued)

Table 2 (continued)

	AI in different transport sectors	AI in transportation systems
	Airport security systems benefit from AI to enhance threat detection. Technologies like facial recognition and AI-based security scanners can identify suspicious behaviors and detect dangerous objects with greater accuracy and speed than traditional methods. This helps prevent security incidents and better manage passenger flows	AI enhances various aspects of airport operations, from security screening to passenger flow management. For example, AI-based facial recognition systems speed up the security screening process, while predictive analytics can optimize staffing levels and resource allocation. This results in a more efficient and pleasant travel experience for passengers
Maritime transport	*Autonomous navigation and ships* (Wang et al. 2023; Wang et al. 2022; Zhang and Zhang 2023; Hwang and Youn 2022; He et al. 2022; Munim et al. 2022; Dai et al. 2024)	
	In the maritime sector, AI is applied in developing autonomous ships. These vessels use advanced sensor systems and AI algorithms to navigate without human intervention	AI is enabling the development of autonomous ships, which can navigate with minimal human intervention. These ships use AI to process data from various sensors, allowing them to avoid obstacles and optimize their routes
	Route optimization and fuel efficiency (Mollaoglu et al. 2023; Li et al. 2022; Öztürk and Başar 2022; Scalia et al. 2023; Zhen et al. 2020)	
	AI algorithms help optimize navigation routes and fuel consumption of ships. By analyzing weather data, ocean currents, and maritime traffic, AI systems can recommend the most efficient routes and adjust ship speed to minimize fuel consumption. This not only reduces costs but also lowers the environmental impact of maritime transportation	AI algorithms help optimize maritime routes by analyzing weather data, ocean currents, and traffic patterns. This leads to more efficient routes and improved fuel efficiency, reducing both costs and environmental impact. AI also assists in managing speed and engine performance to further enhance fuel savings

(continued)

Table 2 (continued)

	AI in different transport sectors	AI in transportation systems
	Predictive maintenance (Makridis et al. 2020; Halvorsen-Weare and Nonås 2023; Angelopoulos et al. 2024; Liu et al. 2022)	
	Predictive maintenance is a key application of AI in maritime transportation as well. Sensors on ships continuously monitor the condition of engines and other critical systems. The collected data is analyzed by machine learning algorithms to anticipate failures and schedule maintenance, improving reliability and reducing the risk of breakdowns at sea	Similar to other transport modes, predictive maintenance is crucial for maritime systems. AI analyzes sensor data on ships to predict equipment failures and schedule maintenance. This proactive approach minimizes downtime and enhances the reliability and safety of maritime operations

- **Autonomous Vehicles**. Two main aspects can be distinguished here:
    1. Self-Driving Cars: Autonomous vehicles (AVs) are at the forefront of AI applications in road transportation. AI algorithms process this data in real time to make driving decisions, such as steering, braking, and accelerating. The promise of Avs includes reduced traffic accidents caused by human error, lower transportation costs, and increased accessibility for individuals unable to drive.
    2. Advanced Driver Assistance Systems (ADAS): While fully autonomous vehicles are still in development, AI-powered ADAS are already enhancing the safety and convenience of driving. Features such as adaptive cruise control, lane-keeping assistance, automatic emergency braking, and parking assistance rely on AI to interpret sensor data and assist drivers in real time. These systems not only improve safety by preventing collisions but also enhance the driving experience.
- **Traffic Management**. When analyzing this aspect, it is important to mention that it is associated with:
    1. Intelligent Traffic Systems: AI is revolutionizing traffic management in urban areas. Intelligent traffic systems use AI to analyze data from traffic cameras, sensors, and GPS devices. These systems can predict traffic congestion, adjust traffic signal timings, and suggest alternative routes to drivers.
    2. Dynamic Traffic Routing: AI-driven dynamic routing applications, such as Google Maps and Waze, provide drivers with real-time traffic information and suggest the fastest routes to their destinations. These applications analyze traffic data, accidents, and road closures to optimize routes, helping reduce travel time and fuel consumption.

Table 3 AI challenges and considerations, own elaboration based on (Athavale et al. 2020; Hong et al. 2023; Bankins et al. 2024; Scheerer and Reussner 2021; Moulahi et al. 2023; Sever and Contissa 2024; Femi Osasona et al. 2024; Khogali and Mekid 2023; Regona et al. 2024; George 2024; Nikitas et al. 2021)

AI in different transport sectors	AI in transportation systems
Safety and reliability (Athavale et al. 2020; Hong et al. 2023; Bankins et al. 2024; Scheerer and Reussner 2021)	
While AI-based systems can potentially improve safety, they also pose new risks. It is crucial to ensure these systems are robust and reliable to avoid failures that could have severe consequences. Validation and certification of AI technologies in critical transportation applications are areas of active research and development	While AI has the potential to significantly enhance transportation systems, it also introduces new safety and reliability challenges. Ensuring that AI systems are robust, fail-safe, and capable of handling unexpected situations is critical. Extensive testing and validation are necessary to certify AI technologies for use in safety–critical applications
Data privacy and security (Moulahi et al. 2023)	
The collection and analysis of large volumes of data are fundamental to the functioning of AI systems. This raises concerns about data privacy and security. It is essential to implement data protection measures to ensure personal and sensitive information is not misused or compromised	The extensive use of data in AI applications raises concerns about privacy and security. Protecting sensitive information and ensuring that data is not misused or compromised is essential. Robust data governance frameworks and security measures must be implemented to address these concerns
Regulation and standards (Sever and Contissa 2024; Osasona et al. 2024; Khogali and Mekid 2023; Regona et al. 2024)	
The adoption of AI in transportation requires an appropriate regulatory framework. It is necessary to develop standards that ensure safety, reliability, and ethics in using AI while fostering innovation. Collaboration between governments, industry, and standardization bodies is crucial to establish these frameworks	The adoption of AI in transportation requires the development of appropriate regulatory frameworks. These frameworks must ensure safety, reliability, and ethical considerations while fostering innovation. Collaboration between governments, industry stakeholders, and regulatory bodies is necessary to create standards supporting the safe and effective deployment of AI in transportation
Impact on employment (George 2024; Nikitas et al. 2021)	
Automation of tasks in the transportation sector could significantly impact employment. While AI creates new job opportunities in areas like developing and maintaining these technologies, it may also lead to job losses in traditional roles. Developing strategies for retraining and reskilling the affected workforce is important	The integration of AI into transportation systems may have significant implications for employment. While AI can create new job opportunities in areas such as AI development, maintenance, and data analysis, it may also lead to job displacement in roles traditionally performed by humans. Addressing this issue requires strategies for workforce retraining and reskilling to help workers adapt to the changing job landscape

- **Smart Infrastructure**. This aspect involves two main processes:
    1. Smart Traffic Lights: AI is being used to develop smart traffic lights that can adapt to changing traffic conditions. These lights use real-time data from cameras and sensors to optimize their timings, reducing idle time at intersections and improving traffic flow. In addition to improving efficiency, smart traffic lights can also enhance safety by prioritizing emergency vehicles and pedestrians.
    2. Connected Vehicles and V2X Communication: Vehicle-to-everything (V2X) communication allows vehicles to communicate with each other and with infrastructure, such as traffic lights and road signs. AI processes the data exchanged in V2X systems to enhance safety and efficiency. For example, connected vehicles can receive alerts about potential hazards, traffic conditions, and upcoming signal changes, enabling them to adjust their speed and driving behaviour accordingly.
- **Predictive Maintenance**:
    1. Vehicle Predictive Maintenance: AI is used to predict and prevent vehicle breakdowns through predictive maintenance. By analyzing data from various sensors monitoring engine performance, tire pressure, and other critical components, AI algorithms can identify patterns that indicate potential issues. This allows for timely maintenance, reducing the risk of sudden breakdowns and extending the lifespan of vehicles. Fleets of commercial vehicles, such as trucks and buses, benefit significantly from predictive maintenance by minimizing downtime and improving operational efficiency.
    2. Infrastructure Predictive Maintenance: AI also plays a crucial role in maintaining road infrastructure. Sensors embedded in roads and bridges collect data on structural integrity, traffic loads, and environmental conditions. AI analyzes this data to predict wear and tear, enabling proactive maintenance. This approach helps prevent infrastructure failures and ensures roads and bridges remain safe and operational.

It is also very important to mention that the application of AI in road transport is focused on safety and security. This important group includes:

1. **Accident Prevention and Response**: AI enhances road safety by preventing accidents and improving response times. AI-powered systems can detect potential collisions and alert drivers or take corrective actions, such as automatic braking. In the event of an accident, AI can assist emergency responders by providing real-time data on the incident location, vehicle conditions, and potential injuries, enabling faster and more effective responses.
2. **Driver Monitoring Systems**: AI-based driver monitoring systems improve safety by detecting signs of driver fatigue, distraction, or impairment. These systems use cameras and sensors to monitor drivers' behaviour and facial expressions. If the AI detects that the driver is not paying attention or is falling asleep, it can issue warnings or take control of the vehicle to prevent accidents.

Considering the challenges of green logistics, it is no less important environmental impact. This effect is expressed through:

1. **Emission Reduction**: AI contributes to reducing the environmental impact of road transportation. Intelligent traffic management systems and dynamic routing applications help reduce congestion and idle times, lowering fuel consumption and emissions. Additionally, AI optimizes the performance of electric vehicles (EVs) by managing battery usage and energy consumption more efficiently.
2. **Eco-Driving Assistance**: AI can also promote eco-friendly driving behaviours. Eco-driving assistance systems analyze driving patterns and provide feedback to drivers on how to reduce fuel consumption and emissions. These systems suggest optimal acceleration, braking, and speed to minimize environmental impact.

However, despite all the advantages and essentials, the application of AI in road transport faces challenges and considerations:

1. **Safety and Reliability**: Ensuring the safety and reliability of AI systems in road transportation is paramount. AI algorithms must be robust and capable of handling diverse and unexpected situations. Rigorous testing, validation, and certification processes are necessary to ensure these systems can operate safely in real-world conditions.
2. **Data Privacy and Security**: The widespread use of AI in road transportation raises concerns about data privacy and security. Protecting the vast amounts of data collected by AI systems, including personal information and real-time location data, is crucial. Strong data governance frameworks and security measures are needed to prevent misuse and ensure user privacy.
3. **Regulatory and Legal Frameworks**: The adoption of AI in road transportation requires appropriate regulatory and legal frameworks. Governments and regulatory bodies must develop standards and guidelines to ensure the safe and ethical use of AI in this sector. This includes addressing liability issues in the event of accidents involving autonomous vehicles and ensuring transparency and accountability in AI decision-making processes.
4. **Social and Economic Impact**: The integration of AI into road transportation may have significant social and economic impacts. While AI can create new job opportunities in areas such as AI development and maintenance, it may also lead to job displacement in traditional roles such as driving. Strategies for workforce retraining and reskilling are essential to help workers adapt to these changes.

In summary, it can be said that the future of AI in road transportation holds immense potential. Advancements in AI technology will continue to drive innovation, leading to more sophisticated and capable systems. Some of the potential future developments include Fully Autonomous Transport Networks, Enhanced V2X Communication, AI-driven urban Planning, and Integration with Other Modes of Transport.

4.3 Recommendations on the Use of AI in Operational Risk Management in Road Transport

The areas of use of AI in road transport identified from the literature, presented in Sect. 4.2, correspond primarily to the needs of risk management at a strategic level (engineering approach). This is because they primarily refer to aspects related to traffic safety based on assessing the risk of adverse events in the form of accidents, congestion, and traffic disruptions, among others. Thanks to AI-based tools, it is possible to take various types of action to reduce the level of risk by diverting traffic to other routes, reducing the speed of moving vehicles and transmitting alerts to drivers to initiate specific preventive activities. AI also plays an important role in mitigating the effects of adverse events that have already occurred. AI tools can quickly trigger emergency procedures and prioritise moving emergency vehicles by interacting with vehicle traffic control systems and relaying alerts. This enables a better response and faster handling of accident victims, as well as damage repair and restoration of traffic after an incident. The use of AI tools to assess the risk of damage to transport infrastructure is also being explored strategically regarding safety. The analytical algorithms developed based on wear patterns make it possible to identify undesirable events in advance, the source of which is damage to infrastructure elements, and to counteract them (recommendations for repairs and maintenance strategies for transport infrastructure).

However, the authors' research area is an interdisciplinary approach to risk assessment in transport systems. Therefore, the identification of the potential for the application of AI in road transport risk management cannot be limited to an engineering approach only. It is also necessary to identify AI-based solutions to support the risk management process at the operational level (process approach). The results of the literature research presented in Sect. 4.2 mainly indicate two solutions that can be used for risk assessment in the transport process:

- Dynamic traffic routing. With AI-driven dynamic routing applications, it is possible to quickly ocean the risk of transporting on a selected route and flexibly modify it to reduce the risk of delay with defined parameters of the transport being carried out. The data from these applications can also form the basis for planning fixed routes for low-risk customer services.
- Vehicle Predictive Maintenance. AI tools can use data from various sources (own historical data, manufacturers' service data) to identify damage patterns in vehicles in use. Combined with data from the sensors with which the vehicles are equipped, it is possible to monitor the vehicle's condition on an ongoing basis, implement an appropriate maintenance strategy and prepare recommendations for drivers on how to use the vehicle.

However, an analysis of scientific publications and industry reports on implementing AI to support managerial decision-making shows the extended potential of applying AI to transport processes. Based on many years of cooperation with road transport companies and the information needs of managers in the process of transport planning and control, well identified in previous research, the potential application of AI tools and the effects of their implementation in terms of risk mitigation at the operational level. The results are presented in Table 4.

5 Conclusions

The research results presented in this article show the increasing popularity of artificial intelligence applications in transport systems, including supporting road transport. The presented area of support for the risk management process in road transport is not closed. The continuing development of technology and the emergence of new AI tools indicate that the needs of managers and decision-makers for implementing such support solutions are constantly evolving.

The road transport literature mainly describes the results of AI applications that support risk management at a strategic level. Therefore, they represent solutions primarily dedicated to risk assessment systems in an engineering approach. However, the analysis of publications on the application of AI in companies from other sectors (e.g. the energy sector) makes it possible to broaden the scope of the use of artificial intelligence in transport processes, which will correspond to the engineering-process approach promoted. Therefore, the results present the potential for implementing AI in risk analysis and proceedings to minimise risks at both the system (engineering) and operational (process) levels.

A limitation of the presented results is how the literature review was conducted. The formulated research questions were answered based on a narrative literature review. Therefore, it is reasonable to extend the scope of the literature review, which will be based on a systematic review of publications.

The results and conclusions provide interesting material for representatives of science and business. The presented results indicate the research potential for analyses focused on the use of AI in creating new risk assessment methods and systems targeted at the specifics of road transport. Managers, on the other hand, may find inspiration for the development of the application of new tools in their decision-making processes at different levels of management. The results obtained have also indicated new research directions for the authors of this article. In the following studies, comparative analyses will be carried out on the application of AI in risk assessment for different transport systems. Research focused on developing risk assessment methods in road transport based on the application of AI is also planned.

Table 4 Distinction between loss and damage as consequences of adverse events, own study

Application area	Effects of implementation	Impact on risk
Analysis of market demand for selected transport services	Improve the transport company's offer and precisely target dedicated customer groups	Reduced risk of lack of interest in the offer
	Better delimitation of service areas and combinations of transports	Reduced risk of missing out on return loads
Analysis of customer behaviour patterns	Better identification of the complex needs of the customer groups served	Reduced risk of inadequate logistics service levels. Lower risk of losing customers
	Optimisation of customer service parameters in terms of costs	Reduced risk of excessive customer service costs
Integration of potential market demand data	Improvement of the transport demand planning process	Reduced risk of inefficient use of the transport fleet
	More accurate planning of reserves, including maintenance processes	Reduced risk of vehicle unavailability
	Improved planning of investments in fleet development	Reduced risk of investment activities that are not economically justified. Lower risk of limited vehicle availability
Analysis of occurring adverse events	Improvement of human resources management, in particular, drivers	Reducing the incidence of human error, incorrect behaviour, and cargo damage
	Improving customer relationship management	Reduced risk of disruptions in communication with customers, reduction of inappropriate customer behaviour. Reduced handling of 'unwanted' customers
	Improved selection of transport routes	Reduced risk of disruptions related to capacity, transport infrastructure, and behaviour of other road users
	Improved freight handling process	Reduced risk of cargo damage during transport, misidentification of data, failure to provide required transport conditions, and lack of required security and documents

References

Aarts L, van Schagen I (2006) Driving speed and the risk of road crashes: A review. Accid Anal Prev 38:215–224. https://doi.org/10.1016/j.aap.2005.07.004

Abdel-Aty M, Ekram AA, Huang H, Choi K (2011) A study on crashes related to visibility obstruction due to fog and smoke. Accid Anal Prev 43:1730–1737. https://doi.org/10.1016/J.AAP.2011.04.003

Ahangar MN, Ahmed QZ, Khan FA, Hafeez M (2021) A survey of autonomous vehicles: enabling communication technologies and challenges. Sensors 21:706. https://doi.org/10.3390/s21030706

Al-Ghamdi AS (2007) Experimental evaluation of fog warning system. Accid Anal Prev 39:1065–1072. https://doi.org/10.1016/J.AAP.2005.05.007

Amundsen FH, Ranes G (2000) Studies on traffic accidents in Norwegian road tunnels. Tunn Undergr Space Technol 15:3–11. https://doi.org/10.1016/S0886-7798(00)00024-9

Angelopoulos A, Giannopoulos A, Nomikos N, Kalafatelis A, Hatziefremidis A, Trakadas P (2024) Federated learning-aided prognostics in the shipping 4.0: principles, workflow, and use cases. IEEE Access 12:6437–6454. https://doi.org/10.1109/ACCESS.2024.3350777

Athavale J, Baldovin A, Graefe R, Paulitsch M, Rosales R (2020) AI and reliability trends in safety-critical autonomous systems on ground and air. In: 2020 50th Annual IEEE/IFIP International Conference on Dependable Systems and Networks Workshops (DSN-W). IEEE, pp 74–77. https://doi.org/10.1109/DSN-W50199.2020.00024

Aydoğan E, Cetek C (2023) Aircraft route optimization with simulated annealing for a mixed airspace composed of free and fixed route structures. Aircr Eng Aerosp Technol 95:637–648. https://doi.org/10.1108/AEAT-11-2021-0343

Aziz S, Dowling MM (2018) AI and machine learning for risk management. SSRN Electron J. https://doi.org/10.2139/ssrn.3201337

Aziz E, Kodur V (2013) An approach for evaluating the residual strength of fire exposed bridge girders. J Constr Steel Res 88:34–42. https://doi.org/10.1016/j.jcsr.2013.04.007

Baker CJ, Reynolds S (1992) Wind-induced accidents of road vehicles. Accid Anal Prev 24:559–575. https://doi.org/10.1016/0001-4575(92)90009-8

Bankins S, Ocampo AC, Marrone M, Restubog SLD, Woo SE (2024) A multilevel review of artificial intelligence in organizations: implications for organizational behavior research and practice. J Organ Behav 45:159–182. https://doi.org/10.1002/job.2735

Bretas AMC, Mendes A, Jackson M, Clement R, Sanhueza C, Chalup S (2023) A decentralised multi-agent system for rail freight traffic management. Ann Oper Res 320:631–661. https://doi.org/10.1007/s10479-021-04178-x

Brijs T, Karlis D, Wets G (2008) Studying the effect of weather conditions on daily crash counts using a discrete time-series model. Accid Anal Prev 40:1180–1190. https://doi.org/10.1016/J.AAP.2008.01.001

British Standard Institution: BS 25999-2:2007—Business continuity management. Specification

Brown ID (1997) Prospects for technological countermeasures against driver fatigue. Accid Anal Prev 29:525–531. https://doi.org/10.1016/S0001-4575(97)00032-8

Budd L, Newstead S (2021) Identifying future vehicle safety priority areas in Australia for the light vehicle fleet. J Road Saf 32:15–24. https://doi.org/10.33492/JRS-D-21-00001

Bureika G, Vaiciunas G, Shi D, Zanuy AC (2022) Influence of track geometry condition monitoring on railway infrastructure maintenance processing. Transp Probl 17:211–220. https://doi.org/10.20858/tp.2022.17.4.18

Carr A, Biswas T, Wheeler JV (2020) Airport operations and security screening: an examination of social justice. J Air Transp Manag 85:101814. https://doi.org/10.1016/j.jairtraman.2020.101814

Cascetta E (2001) Transportation systems engineering: theory and methods. Springer US, Boston, MA. https://doi.org/10.1007/978-1-4757-6873-2

Ćelić J, Mandžuka B, Tomas V, Tadić F (2024) Driver-centric urban route planning: smart search for parking. Sustainability 16:856. https://doi.org/10.3390/su16020856

Chen C, Zhu L, Zhao H (2022) 5G-enabled edge intelligence for autonomous train control: a practical perspective. In: 2022 IEEE 25th International Conference on Intelligent Transportation Systems (ITSC). IEEE, pp 1790–1794. https://doi.org/10.1109/ITSC55140.2022.9922178

Cheng J, Ju M, Zhou M, Liu C, Gao S, Abusorrah A, Jiang C (2022) A dynamic evolution method for autonomous vehicle groups in a highway scene. IEEE Internet Things J 9:1445–1457. https://doi.org/10.1109/JIOT.2021.3086832

Clarke J, Obrien E (2016) A Multi-hazard risk assessment methodology, stress test framework and decision support tool for transport infrastructure networks. Transp Res Procedia 14:1355–1363. https://doi.org/10.1016/J.TRPRO.2016.05.208

Cuppi F, Vignali V, Lantieri C, Rapagnà L, Dimola N, Galasso T (2021) High density european rail traffic management system (HD-ERTMS) for urban railway nodes: the case study of Rome. J Rail Transp Plan Manag 17:100232. https://doi.org/10.1016/j.jrtpm.2020.100232

Dai Y, He Y, Huang L, Zhao X (2024) Autonomous navigation decision-making for ships in complex estuarine waters: methodology and validation. Ocean Eng 309:118474. https://doi.org/10.1016/j.oceaneng.2024.118474

Dauptain X, Koné A, Grolleau D, Cerezo V, Gennesseaux M, Do M-T (2022) Conception of a high-level perception and localization system for autonomous driving. Sensors 22:9661. https://doi.org/10.3390/s22249661

de Pater I, Reijns A, Mitici M (2022) Alarm-based predictive maintenance scheduling for aircraft engines with imperfect remaining useful life prognostics. Reliab Eng Syst Saf 221:108341. https://doi.org/10.1016/j.ress.2022.108341

Deng Z, Song H, Huang H, Li Y, Dong H (2020) Multi-sensor based train localization and data fusion in autonomous train control system. In: 2020 Chinese Automation Congress (CAC). IEEE, pp 5702–5707. https://doi.org/10.1109/CAC51589.2020.9327825

Di Vaio A, Palladino R, Hassan R, Escobar O (2020) Artificial intelligence and business models in the sustainable development goals perspective: A systematic literature review. J Bus Res 121:283–314. https://doi.org/10.1016/j.jbusres.2020.08.019

Ding S, Zhang Y, Duygun M (2019) Modeling price volatility based on a genetic programming approach. Br J Manag 30:328–340. https://doi.org/10.1111/1467-8551.12359

Doecke SD, Kloeden CN, Dutschke JK, Baldock MRJ (2018) Safe speed limits for a safe system: The relationship between speed limit and fatal crash rate for different crash types. Traffic Inj Prev 19:404–408. https://doi.org/10.1080/15389588.2017.1422601

Dong Y, Frangopol DM, Saydam D (2013) Time-variant sustainability assessment of seismically vulnerable bridges subjected to multiple hazards. Earthq Eng Struct Dyn 42:1451–1467. https://doi.org/10.1002/eqe.2281

Dong Y, Frangopol DM, Saydam D (2014) Sustainability of highway bridge networks under Seismic Hazard. J Earthquake Eng 18:41–66. https://doi.org/10.1080/13632469.2013.841600

Du X, Zhao X, Wu C-H, Feng K (2021) Functionality, emotion, and acceptance of artificial intelligence virtual assistants. J Glob Inf Manag 30:1–21. https://doi.org/10.4018/JGIM.290418

Dwivedi YK, Hughes L, Ismagilova E, Aarts G, Coombs C, Crick T, Duan Y, Dwivedi R, Edwards J, Eirug A, Galanos V, Ilavarasan PV, Janssen M, Jones P, Kar AK, Kizgin H, Kronemann B, Lal B, Lucini B, Medaglia R, Le Meunier-FitzHugh K, Le Meunier-FitzHugh LC, Misra S, Mogaji E, Sharma SK, Singh JB, Raghavan V, Raman R, Rana NP, Samothrakis S, Spencer J, Tamilmani K, Tubadji A, Walton P, Williams MD (2021) Artificial Intelligence (AI): Multidisciplinary perspectives on emerging challenges, opportunities, and agenda for research, practice and policy. Int J Inf Manage 57:101994. https://doi.org/10.1016/j.ijinfomgt.2019.08.002

Edwards JS, Duan Y, Robins PC (2000) An analysis of expert systems for business decision making at different levels and in different roles. Eur J Inf Syst 9:36–46. https://doi.org/10.1057/palgrave.ejis.3000344

Ellefsen APT, Oleśków-Szłapka J, Pawłowski G, Toboła A (2019) Striving for excellence in AI implementation: AI maturity model framework and preliminary research results. Logforum 15:363–376. https://doi.org/10.17270/J.LOG.2019.354

Institute of risk management: From the cube to the rainbow double helix: a risk practitioner's guide to the COSO ERM Frameworks, London

Eskenazi AG, Joshi AP, Butler LG, Ryerson MS (2023) Equitable optimization of US airline route networks. Comput Environ Urban Syst 102:101973. https://doi.org/10.1016/j.compenvurbsys.2023.101973

Ettema D, Friman M, Olsson LE, Gärling T (2017) Season and weather effects on travel-related mood and travel satisfaction. Front Psychol 8. https://doi.org/10.3389/fpsyg.2017.00140

Fritz-Morgenthal S, Hein B, Papenbrock J (2022) Financial risk management and explainable, trustworthy, responsible AI. Front Artif Intell 5. https://doi.org/10.3389/frai.2022.779799

Garbuio M, Lin N (2019) Artificial intelligence as a growth engine for health care startups: Emerging business models. Calif Manage Rev 61:59–83. https://doi.org/10.1177/0008125618811931

Garikapati D, Shetiya SS (2024) Autonomous vehicles: evolution of artificial intelligence and the current industry landscape. Big Data Cogn Comput 8:42. https://doi.org/10.3390/bdcc8040042

Garmabaki AHS, Thaduri A, Famurewa S, Kumar U (2021) Adapting railway maintenance to climate change. Sustainability 13:13856. https://doi.org/10.3390/su132413856

George AS (2024) Artificial intelligence and the future of work: job shifting not job loss. Partners Universal Innovative Research Publication, vol 2, pp 17–37

Gevaert CM, Carman M, Rosman B, Georgiadou Y, Soden R (2021) Fairness and accountability of AI in disaster risk management: Opportunities and challenges. Patterns 2:100363. https://doi.org/10.1016/j.patter.2021.100363

Govindan K, Jafarian A, Nourbakhsh V (2019) Designing a sustainable supply chain network integrated with vehicle routing: A comparison of hybrid swarm intelligence metaheuristics. Comput Oper Res 110:220–235. https://doi.org/10.1016/j.cor.2018.11.013

Gu K, Hu J, Jia W (2023) Adaptive area-based traffic congestion control and management scheme based on fog computing. IEEE Trans Intell Transp Syst 24:1359–1373. https://doi.org/10.1109/TITS.2022.3183687

Guo Y, Wang Y, Khan F, Al-Atawi AA, Al Abdulwahid A, Lee Y, Marapelli B (2023) Traffic management in IoT backbone networks using GNN and MAB with SDN orchestration. Sensors 23:7091. https://doi.org/10.3390/s23167091

Gursel E, Madadi M, Coble JB, Agarwal V, Yadav V, Boring RL, Khojandi A (2025) The role of AI in detecting and mitigating human errors in safety-critical industries: A review. Reliab Eng Syst Saf 256:110682. https://doi.org/10.1016/j.ress.2024.110682

Haight FA (1986) Risk, especially risk of traffic accident. Accid Anal Prev 18:359–366. https://doi.org/10.1016/0001-4575(86)90009-6

Halvorsen-Weare EE, Nonås LM (2023) Maritime logistics optimisation for predictive maintenance at offshore wind farms. J Phys Conf Ser 2626:012040. https://doi.org/10.1088/1742-6596/2626/1/012040

Hauer E (1982) Traffic conflicts and exposure. Accid Anal Prev 14:359–364. https://doi.org/10.1016/0001-4575(82)90014-8

He Y, Liu X, Zhang K, Mou J, Liang Y, Zhao X, Wang B, Huang L (2022) Dynamic adaptive intelligent navigation decision making method for multi-object situation in open water. Ocean Eng 253:111238. https://doi.org/10.1016/j.oceaneng.2022.111238

Heim S, Clemens J, Steck JE, Basic C, Timmons D, Zwiener K (2020) Predictive maintenance on aircraft and applications with digital twin. In: 2020 IEEE International Conference on Big Data (Big Data). IEEE, pp 4122–4127. https://doi.org/10.1109/BigData50022.2020.9378433

Herrmann T, Pfeiffer S (2023) Keeping the organization in the loop: a socio-technical extension of human-centered artificial intelligence. AI Soc 38:1523–1542. https://doi.org/10.1007/s00146-022-01391-5

Hildemann M, Verstegen JA (2023) 3D-flight route optimization for air-taxis in urban areas with evolutionary algorithms and GIS. J Air Transp Manag 107:102356. https://doi.org/10.1016/j.jairtraman.2022.102356

Hong Y, Lian J, Xu L, Min J, Wang Y, Freeman LJ, Deng X (2023) Statistical perspectives on reliability of artificial intelligence systems. Qual Eng 35:56–78. https://doi.org/10.1080/08982112.2022.2089854

Hossain MA, Akter S, Yanamandram V, Gunasekaran A (2022) Operationalizing Artificial Intelligence-Enabled customer analytics capability in retailing. J Glob Inf Manag 30:1–23. https://doi.org/10.4018/JGIM.298992

https://www.investopedia.com/terms/t/transportation_sector.asp

Hussain Q, Feng H, Grzebieta R, Brijs T, Olivier J (2019) The relationship between impact speed and the probability of pedestrian fatality during a vehicle-pedestrian crash: A systematic review and meta-analysis. Accid Anal Prev 129:241–249. https://doi.org/10.1016/j.aap.2019.05.033

Hwang T, Youn I-H (2022) Navigation scenario permutation model for training of maritime autonomous surface ship remote operators. Appl Sci 12:1651. https://doi.org/10.3390/app12031651

ISO 31000:2018 Risk management—Guidelines

ISO 31010:2019 Risk management—Risk assessment techniques

Jain S, Ahuja NJ, Srikanth P, Bhadane KV, Nagaiah B, Kumar A, Konstantinou C (2021) Blockchain and autonomous vehicles: recent advances and future directions. IEEE Access 9:130264–130328. https://doi.org/10.1109/ACCESS.2021.3113649

Janssen S, van der Sommen R, Dilweg A, Sharpanskykh A (2020) Data-driven analysis of airport security checkpoint operations. Aerospace 7:69. https://doi.org/10.3390/aerospace7060069

Jarrahi MH (2018) Artificial intelligence and the future of work: Human-AI symbiosis in organizational decision making. Bus Horiz 61:577–586. https://doi.org/10.1016/j.bushor.2018.03.007

Jun W, Son M, Yoo J, Lee S (2023) Optimal configuration of multi-task learning for autonomous driving. Sensors 23:9729. https://doi.org/10.3390/s23249729

Kaplan A, Haenlein M (2019) Siri, Siri, in my hand: Who's the fairest in the land? On the interpretations, illustrations, and implications of artificial intelligence. Bus Horiz 62:15–25. https://doi.org/10.1016/j.bushor.2018.08.004

Karamitsos I, Papadaki M, Al-Hussaeni K, Kanavos A (2023) Transforming airport security: enhancing efficiency through blockchain smart contracts. Electronics (Basel) 12:4492. https://doi.org/10.3390/electronics12214492

Khogali HO, Mekid S (2023) The blended future of automation and AI: examining some long-term societal and ethical impact features. Technol Soc 73:102232. https://doi.org/10.1016/j.techsoc.2023.102232

Kowalik B, Szpyrka M (2019) An Entropy-Based car failure detection method based on data acquisition pipeline. Entropy 21:426. https://doi.org/10.3390/e21040426

Kumar V, Sharma P, Chinnappa Naidu R, Kaleem Khodabux M, Sewraj K, Anand Mohajeer A (2022) Airplane flight route optimization problem with multi-constraints. In: 2022 Second International Conference on Advances in Electrical, Computing, Communication and Sustainable Technologies (ICAECT). IEEE, pp 1–6. https://doi.org/10.1109/ICAECT54875.2022.9807926

Kuo T-C, Smith S (2018) A systematic review of technologies involving eco-innovation for enterprises moving towards sustainability. J Clean Prod 192:207–220. https://doi.org/10.1016/j.jclepro.2018.04.212

Kurek A, Macioszek E (2021) Impact of parking Maneuvers on the capacity of the inlets of intersections with traffic lights for road traffic conditions in poland. Sustainability 14:432. https://doi.org/10.3390/su14010432

La Scalia G, Mancini S, Adelfio L, Giallanza A (2023) Development of a vessel scheduling optimization model to improve maritime transport sustainability. Sustain Futur 6:100123. https://doi.org/10.1016/j.sftr.2023.100123

Lee J, Mitici M (2022) Multi-objective design of aircraft maintenance using Gaussian process learning and adaptive sampling. Reliab Eng Syst Saf 218:108123. https://doi.org/10.1016/j.ress.2021.108123

Lee J, Mitici M, Blom HAP, Bieber P, Freeman F (2023) Analyzing emerging challenges for data-driven predictive aircraft maintenance using agent-based modeling and hazard identification. Aerospace 10:186. https://doi.org/10.3390/aerospace10020186

Levine N, Kim KE, Nitz LH (1995) Daily fluctuations in Honolulu motor vehicle accidents. Accid Anal Prev 27:785–796. https://doi.org/10.1016/0001-4575(95)00038-0

Li M, Shang X, Liu N, Pan X, Han F (2022) Knowledge management in relationship among abusive management, self-efficacy, and corporate performance under artificial intelligence. J Glob Inf Manag 30:1–26. https://doi.org/10.4018/JGIM.307067

Li X, Sun B, Jin J, Ding J (2022) Speed optimization of container ship considering route segmentation and weather data loading: turning point-time segmentation method. J Mar Sci Eng 10:1835. https://doi.org/10.3390/jmse10121835

Li Y, Zhu L (2022) Collaborative cloud and edge computing in 5G based train control systems. In: GLOBECOM 2022—2022 IEEE Global Communications Conference. IEEE, pp 2542–2547. https://doi.org/10.1109/GLOBECOM48099.2022.10001291

Li L, Song H, Ma J, Dong H (2022) Description and analysis of train-centric communication based autonomous train control system. In: 2022 IEEE 25th International Conference on Intelligent Transportation Systems (ITSC). IEEE, pp 3315–3320. https://doi.org/10.1109/ITSC55140.2022.9922576

Liu Q, Li J (2022) The progress of business analytics and knowledge management for enterprise performance using artificial intelligence and Man-Machine coordination. J Glob Inf Manag 30:1–21. https://doi.org/10.4018/JGIM.302642

Liu S, Chen H, Shang B, Papanikolaou A (2022) Supporting predictive maintenance of a ship by analysis of onboard measurements. J Mar Sci Eng 10:215. https://doi.org/10.3390/jmse10020215

Liu Z, Coleman N, Patrascu FI, Yin K, Li X, Mostafavi A (2025) Artificial intelligence for flood risk management: A comprehensive state-of-the-art review and future directions. Int J Disaster Risk Reduct 117:105110. https://doi.org/10.1016/j.ijdrr.2024.105110

Liu X, Ma X, Shan B (2023) Value creation and management improvement of AI in the power industry. In: 2023 International Conference on Integrated Intelligence and Communication Systems (ICIICS). IEEE, pp 1–6. https://doi.org/10.1109/ICIICS59993.2023.10421492

Luo J, Peng Q, Wen C, Wen W, Huang P (2022) Data-driven decision support for rail traffic control: a predictive approach. Expert Syst Appl 207:118050. https://doi.org/10.1016/j.eswa.2022.118050

Mahroof K (2019) A human-centric perspective exploring the readiness towards smart warehousing: The case of a large retail distribution warehouse. Int J Inf Manage 45:176–190. https://doi.org/10.1016/j.ijinfomgt.2018.11.008

Makarius EE, Mukherjee D, Fox JD, Fox AK (2020) Rising with the machines: A sociotechnical framework for bringing artificial intelligence into the organization. J Bus Res 120:262–273. https://doi.org/10.1016/j.jbusres.2020.07.045

Makridis G, Kyriazis D, Plitsos S (2020) Predictive maintenance leveraging machine learning for time-series forecasting in the maritime industry. In: 2020 IEEE 23rd International Conference on Intelligent Transportation Systems (ITSC). IEEE, pp 1–8. https://doi.org/10.1109/ITSC45102.2020.9294450

Manzoor S, Jamil H, Nawaz M (2024) Technology integration in ESL classroom: advantages and challenges. Int J Lang Lit Transl 7. https://doi.org/10.36777/ijollt2024.7.1.102

Mollaoglu M, Altay BC, Balin A (2023) Bibliometric review of route optimization in maritime transportation: environmental sustainability and operational efficiency. Transp Res Rec: J Transp Res Board 2677:879–890. https://doi.org/10.1177/03611981221150922

Moulahi T, Jabbar R, Alabdulatif A, Abbas S, El Khediri S, Zidi S, Rizwan M (2023) Privacy-preserving federated learning cyber-threat detection for intelligent transport systems with blockchain-based security. Expert Syst 40. https://doi.org/10.1111/exsy.13103

Munim ZH, Saha R, Schøyen H, Ng AKY, Notteboom TE (2022) Autonomous ships for container shipping in the Arctic routes. J Mar Sci Technol 27:320–334. https://doi.org/10.1007/s00773-021-00836-8

Nagy J, Foltin P, Ondryhal V (2022) Use of big data analysis to identify possible sources of supply chain disruption through the DOTMLPFI method. Logforum 18:309–319. https://doi.org/10.17270/J.LOG.2022.731

Naji M, Braytee A, Anaissi A, Sianaki OA, Al-Ani A (2019) Optimizing the waiting time for airport security screening using multiple queues and servers. In Complex, Intelligent, and Software Intensive Systems: Proceedings of the 13th International Conference on Complex, Intelligent, and Software Intensive Systems (CISIS-2019). Springer International Publishing, pp 496-507. https://doi.org/10.1007/978-3-030-22354-0_44

Niewczas A, Mórawski Ł, Rymarz J, Dębicka E, Hołyszko P (2023) Operational risk assessment model for city buses. Eksploatacja i Niezawodność—Maintenance and Reliability. https://doi.org/10.17531/ein.2023.1.14

Nikitas A, Vitel A-E, Cotet C (2021) Autonomous vehicles and employment: an urban futures revolution or catastrophe? Cities 114:103203. https://doi.org/10.1016/j.cities.2021.103203

Oron-Gilad T, Shinar D (2000) Driver fatigue among military truck drivers. Transp Res Part F Traffic Psychol Behav 3:195–209. https://doi.org/10.1016/S1369-8478(01)00004-3

Osasona F, Amoo OO, Atadoga A, Abrahams TO, Farayola OA, Ayinla B (2024) Reviewing the ethical implications of ai in decision making processes. Int J Manag Entrep Research 6:322–335. https://doi.org/10.51594/ijmer.v6i2.773

Østerlund C, Jarrahi MH, Willis M, Boyd KT, Wolf C (2021) Artificial intelligence and the world of work, a co-constitutive relationship. J Assoc Inf Sci Technol 72:128–135. https://doi.org/10.1002/asi.24388

Öztürk OB, Başar E (2022) Multiple linear regression analysis and artificial neural networks based decision support system for energy efficiency in shipping. Ocean Eng 243:110209. https://doi.org/10.1016/j.oceaneng.2021.110209

Pipicelli M, Gimelli A, Sessa B, De Nola F, Toscano G, Di Blasio G (2024) Architecture and potential of connected and autonomous vehicles. Vehicles 6:275–304. https://doi.org/10.3390/vehicles6010012

Qiao J, Zhang D, de Jonge D (2022) Priority-based traffic management protocols for autonomous vehicles on road networks. Presented at the Australasian Joint Conference on Artificial Intelligence. Springer International Publishing, pp 240–253. https://doi.org/10.1007/978-3-030-97546-3_20

Ravish R, Swamy SR (2021) Intelligent traffic management: a review of challenges, solutions, and future perspectives. Transp Telecommun J 22:163–182. https://doi.org/10.2478/ttj-2021-0013

Regona M, Yigitcanlar T, Hon C, Teo M (2024) Artificial intelligence and sustainable development goals: Systematic literature review of the construction industry. Sustain Cities Soc 108:105499. https://doi.org/10.1016/j.scs.2024.105499

Rojek I, Jasiulewicz-Kaczmarek M, Piechowski M, Mikołajewski D (2023) An artificial intelligence approach for improving maintenance to supervise machine failures and support their repair. Appl Sci 13:4971. https://doi.org/10.3390/app13084971

Russell SJ, Norvig P (2016) Artificial intelligence: A modern approach. Pearson

Sagberg F (1999) Road accidents caused by drivers falling asleep. Accid Anal Prev 31:639–649. https://doi.org/10.1016/S0001-4575(99)00023-8

Saleem AA, Siddiqui HUR, Shafique R, Haider A, Ali M (2020) A review on smart IOT based parking system. In Recent Advances on Soft Computing and Data Mining: Proceedings of the Fourth International Conference on Soft Computing and Data Mining (SCDM 2020). Springer International Publishing, Melaka, Malaysia, pp 264-273. https://doi.org/10.1007/978-3-030-36056-6_26

Satterthwaite SP (1976) An assessment of seasonal and weather effects on the frequency of road accidents in California. Accid Anal Prev 8:87–96. https://doi.org/10.1016/0001-4575(76)90002-6

Scheerer M, Reussner R (2021) Reliability prediction of self-adaptive systems managing uncertain AI black-box components. In: 2021 International Symposium on Software Engineering for

Adaptive and Self-Managing Systems (SEAMS). IEEE, pp 111–117. https://doi.org/10.1109/SEAMS51251.2021.00024

Schneider S, Leyer M (2019) Me or information technology? Adoption of artificial intelligence in the delegation of personal strategic decisions. Manag Decis Econ 40:223–231. https://doi.org/10.1002/mde.2982

Seeber I, Bittner E, Briggs RO, de Vreede T, de Vreede G-J, Elkins A, Maier R, Merz AB, Oeste-Reiß S, Randrup N, Schwabe G, Söllner M (2020) Machines as teammates: A research agenda on AI in team collaboration. Inf Manag 57:103174. https://doi.org/10.1016/j.im.2019.103174

Sever T, Contissa G (2024) Automated driving regulations—where are we now? Transp Res Interdiscip Perspect. 24:101033. https://doi.org/10.1016/j.trip.2024.101033

Smalko Z, Nowakowski T, Tubis A (2020) Introduction to reliability based hazard theory (in Polish). Wroclaw University of Science and Technology Publishing House, Wrocław

Smoczyński P, Gill A, Kadziński A (2021) Maintenance layers for railway infrastructure in Poland. Transport 35:605–615. https://doi.org/10.3846/transport.2020.14137

Song H, Gao S, Li Y, Liu L, Dong H (2023) Train-centric communication based autonomous train control system. IEEE Trans Intell Veh 8:721–731. https://doi.org/10.1109/TIV.2022.3192476

Stanton I, Munir K, Ikram A, El-Bakry M (2023) Predictive maintenance analytics and implementation for aircraft: challenges and opportunities. Syst Eng 26:216–237. https://doi.org/10.1002/sys.21651

Stoilova S (2022) A new aggregated multi-criteria approach for evaluation of the autonomous metro systems performance in the European Countries. Symmetry (Basel). 14. https://doi.org/10.3390/sym14102025

Stopka O, Stopkova M, Ľuptak V, Krile S (2020) Application of the chosen multi-criteria decision—making methods to identify the autonomous train system supplier. Transp Probl 15:45–57. https://doi.org/10.21307/tp-2020-019

Szabo S, Blistanova M, Mako S, Vajdova I, Pilat M (2020) Algorithmic optimization of the calculation with the consideration of the interconnection of the basic economic parameters of the flight route of the model air carrier. Manag Syst Prod Eng 28:34–39. https://doi.org/10.2478/mspe-2020-0006

Tuan MND, Thanh NN, Le Tuan L (2019) Applying a mindfulness-based reliability strategy to the internet of things in healthcare—A business model in the Vietnamese market. Technol Forecast Soc Change 140:54–68. https://doi.org/10.1016/j.techfore.2018.10.024

Tubis A (2018) Operational risk management method in road transport. Oficyna Wydawnicza Politechniki Wrocławskiej, Wrocław

Van Gheluwe C, Semanjski I, Hendrikse S, Gautama S (2020) Geospatial dashboards for intelligent multimodal traffic management. In: 2020 IEEE International Conference on Pervasive Computing and Communications Workshops (PerCom Workshops). IEEE, pp 1–3. https://doi.org/10.1109/PerComWorkshops48775.2020.9156231

Vuilleumier F, Weatherill A, Crausaz B (2002) Safety aspects of railway and road tunnel: Example of the Lötschberg railway tunnel and Mont-blanc road tunnel. Tunn Undergr Space Technol 17:153–158. https://doi.org/10.1016/S0886-7798(02)00018-4

Wang Y-M, Liu J, Elhag TMS (2008) An integrated AHP–DEA methodology for bridge risk assessment. Comput Ind Eng 54:513–525. https://doi.org/10.1016/j.cie.2007.09.002

Wang S, Zhang Y, Zhang X, Gao Z (2023) A novel maritime autonomous navigation decision-making system: modeling, integration, and real ship trial. Expert Syst Appl 222:119825. https://doi.org/10.1016/j.eswa.2023.119825

Wang S, Zhang, Y, Wang X, Huo R, Song F, Zhai P (2022) A Novel maritime autonomous navigation system: from design to real ship trial. In: 2022 IEEE 25th International Conference on Intelligent Transportation Systems (ITSC). IEEE, pp 1362–1366. https://doi.org/10.1109/ITSC55140.2022.9921769

Williamson AM, Feyer A-M (2000) Moderate sleep deprivation produces impairments in cognitive and motor performance equivalent to legally prescribed levels of alcohol intoxication. Occup Environ Med 57:649–655. https://doi.org/10.1136/oem.57.10.649

Zhang J, Zhang J (2023) Artificial intelligence applied on traffic planning and management for rail transport: a review and perspective. Discrete Dyn Nat Soc. 2023:1–17. https://doi.org/10.1155/2023/1832501

Zhang W, Zhang Y (2023) Navigation risk assessment of autonomous ships based on entropy–TOPSIS–coupling coordination model. J Mar Sci Eng. 11:422. https://doi.org/10.3390/jmse11020422

Zhen L, Hu Z, Yan R, Zhuge D, Wang S (2020) Route and speed optimization for liner ships under emission control policies. Transp Res Part C Emerg Technol 110:330–345. https://doi.org/10.1016/j.trc.2019.11.004

Zheng Q, Gou J, Camarinha-Matos LM, Zhang JZ, Zhang X (2023) Digital capability requirements and improvement strategies: Organizational socialization of AI teammates. Inf Process Manag 60:103504. https://doi.org/10.1016/j.ipm.2023.103504

Reverse Logistics in E-commerce—Case Study of Croatia

Petra Paunović, Maja Trstenjak, Goran Đukić, and Tihomir Opetuk

Abstract Reverse logistics processes and related tools can be complex to implement optimally, so the users face numerous challenges in daily activities. Reverse logistics in an e-commerce in most cases is related to return or exchange logistics, specifically the process where customers return goods that do not conform to the order requirements. This is one of the most important steps in the productive management of relationships with the customer which affects their loyalty, which is why the development of reverse logistics is imperative in e-commerce. In this paper, the research was conducted with the representatives of companies who sell their products through webshops, but also with the customers to get an insight into the opinions and attitudes of companies and the related habits of customers regarding the return and ordering of products. The results showed that to optimize return logistics, it is necessary to integrate an approach that includes technological solutions as well as to build an exceptionally good communication channel between customers and stores. Another key factor for the success of a web store is cooperation with reliable logistics partners. When a company cooperates with such partners, there is the possibility of tracking shipments and clear guidelines for returns, difficulties are significantly reduced, and then both customers and retailers themselves are more satisfied. Investing in the improvement of return logistics will not only contribute to customer satisfaction, but ultimately bring long-term profits and competitiveness of the company on the market.

Keywords Logistics · Reverse logistics · E-commerce · Return logistics

P. Paunović · M. Trstenjak (✉) · G. Đukić · T. Opetuk
Faculty of Mechanical Engineering, University of Zagreb, Zagreb, Croatia
e-mail: maja.trstenjak@fsb.unizg.hr

P. Paunović
e-mail: pp211981@stud.fsb.hr

G. Đukić
e-mail: goran.dukic@fsb.unizg.hr

T. Opetuk
e-mail: tihomir.opetuk@fsb.unizg.hr

A. Kolinski and M. Adamczak (eds.), *Digitalisation of the Greening Supply Chain*, EcoProduction, https://doi.org/10.1007/978-3-031-88918-9_3

1 Introduction

Nowadays, there is hardly a single company that does not have its website. In addition to general information about the company that we can find on the web, such as what the company does, websites often contain e-stores that allow users to shop without going to a physical store, from the comfort of their own home. Lockdown in 2020 and the pandemic of the Covid-19 virus only contributed to this practice, so even people who were not used to it until then adapted to the use of e-shops. This greatly increased the demand for e-commerce, and many companies that did not have such an option until then had to seriously consider this approach if they wanted to remain competitive in the market. However, there are frequent situations when customers are not satisfied with the ordered product, the product does not look like the pictures, the delivered product is damaged, or the product simply does not fit for some other reason. Then the ordered goods are returned to the seller. Although the customer can easily return the ordered item, for the seller the whole process is not so simple, and thus the importance of return logistics increases.

The need to introduce return logistics flows arose mainly due to the demand for a different approach in managing logistics flows, which is determined by an increasingly demanding market, but also due to the accompanying environmental impacts. Return logistics mainly refers to the process in which customers return goods that do not conform to the requirements of the order (Waters 2003). The main goal of reverse logistics is to optimize or improve efficiency after-sales activities. In addition, customer satisfaction is also important, as is the fact that certain products are given a new application after use, that is, that the products do not end up as waste. Material recycling does not only have the advantage of preserving the environment, but also offers economic benefits, and it is precisely because of the advantages that return logistics has developed significantly in recent years and is becoming interesting to an increasing number of people and companies.

The term reverse logistics began to be used several decades ago, but the exact moment of its first use cannot be determined. Return channels or flows as a term began to appear in the scientific literature during the 70s of the 20th century and mainly refer to recycling activities. Lambert and Stock (1981) at the beginning of the 80s of the 20th century interpreted reverse logistics flows as a wrong path because most shipments move from the producer to the customer. Murphy and Poist (1989) also agree with this interpretation and define reverse logistics as the process of movement of goods from the consumer to the producer through the distribution channel.

From the aforementioned interpretations, it is evident that in the initial period, the significance of return logistics flows was not sufficiently recognized and was limited only to some specific activities, so return flows were considered unusual and were very rare.

The first formal definition of reverse logistics was given by the Council for Supply Chain Management Professionals (CSCMP) in the early 1990s (Aćimović et al. 2020). According to the Council, the concept of reverse logistics refers to the role of logistics in recycling, waste disposal and hazardous materials management. The broader description of this term includes other logistics activities related to the reduction of used resources, exchange, reuse and disposal of materials. However, according to today's criteria, this definition is still too general and imprecise.

In the period of the 90s of the 20th century, there was a more intense interest in the activities of reverse logistics and the advantages it provides were seen, and therefore the definition became wider, developing the range of activities and entities involved in the process itself (Aćimović et al. 2020).

1.1 Importance of Reverse Logistics as a Starting Point for Green Logistics

Reverse logistics is an important starting point for green logistics due to its overall operation. Currently, Green Supply Chain Management (GSCM) is gaining more and more importance both in theory and in practice, combining environmental management and supply chain management (SCM). GSCM represents the integration of ecological thinking in SCM activities (Patak et al. 2020; Ivanova et al. 2022). Building a green supply chain (GSC) requires the application of several green practices. These are environmentally friendly activities that reduce the impact on the environment and realize the added value of these activities. At the same time, they contribute to lean production and enable the company to maintain a competitive advantage. Most of these practices are applicable at the level of one company, however, some require the cooperation of several partners in the chain. Environmentally proactive manufacturers directly and positively influence the environmental behavior of their business partners and implement internal green practices with extension to external partners in the supply chain, thus increasing the chance for its successful implementation. Greater or lesser involvement of business partners in common goals includes setting, coordinating, and achieving environmental goals that will lead to a reduction in environmental impact. Developing external green cooperation significantly improves a company's environmental performance.

One of the green practices that requires a greater or lesser degree of cooperation between business partners in the chain is reverse logistics. The return logistics cycle manages the packaging and return of defective or rejected products. However, reverse logistics should be considered throughout the entire product life cycle, including the product design part. The basic life cycle of a product includes the following stages: concept, design, production, use, recovery, recycling and/or disposal in landfills (Patak et al. 2020). The implementation of a reverse logistics network is a strategic decision aimed at minimizing costs, maximizing profits, customer satisfaction and environmental benefits.

As for return logistics research, it have been carried out mainly since 1992 when Stock (1992) recognized reverse logistics as a topic of social interest. The interest of researchers mainly is focused on the strategic management of reverse logistics, especially the application of methods and approaches to reverse logistics management, the implementation of reverse logistics systems and the development of cooperation between business partners in the supply chain, especially with third-party logistics partners.

In a study that focused on the cooperation of supply chain partners in the implementation of green practices (including reverse logistics), it was found that environmental cooperation and monitoring between customers and their suppliers improves environmental and business performance. Moreover, Gimenez and Sierra (2013) found that the effect on environmental performance is positive and can also be improved by assessing suppliers' environment. A positive relationship between green practices, green supply chain and environmental (and economic) performance was also found in food retail research (Petljak et al. 2018).

Dangelico and Pontrandolfo (2015) determined that the ability to develop ecological cooperation and carry out ecological activities significantly affects business performance.

Yen (2018) deals with the internal and external antecedents of the application of green practices, specifically which of them affect the cooperation between the buyer and the supplier and how. It develops the results of previous research (Caniëls et al. 2013) which suggest that the main drivers of supplier involvement in green activities are supplier readiness and customer requirements. The role of customers in the application of green activities was also confirmed in the research of Chavez et al. (2016). They found that customer pressure has a significant impact on the implementation of green initiatives, which in turn leads to a multifold increase in operational performance. However, so far no one has investigated the extent to which customers' requests for reverse logistics services contribute to building a green supply chain. Therefore, primary research was conducted based on the fact that if requests for return logistics services prove to be important, suppliers will be motivated to fulfill them. This creates an opportunity to build reverse logistics systems. Assuming that reverse logistics is a key activity in building a green supply chain, the importance of reverse logistics services will be directly related to building a green supply chain (Patak et al. 2020).

2 Return Logistics in an E-Trade Environment

Returns logistics in an e-commerce environment mainly refers to processes where customers return goods that do not conform to the order requirements. When the products purchased online are found to be of inferior quality or unsatisfactory in other respects, they will be returned to the suppliers or manufacturers according to their refund or exchange requests from the retailers. The specific return logistics flow should include logistics flow, information flow, capital flow, and all or part of the business flow. Due to the current ignorance of most retailers about return logistics, the return or exchange issue is always one of the biggest contradictions between retailers and customers. Therefore, the development of return logistics is imperative in e-commerce (Caniëls et al. 2013).

An additional support for the increasing use of return logistics is the legal regulation, because according to EU rules, the trader must repair or replace the goods, lower the price or refund the money if the purchased goods are defective, do not look or function as shown in the ad. If the product or service was purchased online or outside the store (by phone, mail order or from a traveling salesman), the customer has the right to cancel and return the order within 14 days, for any reason and without explanation (Chavez et al. 2016).

2.1 *Characteristics of Return Logistics in e-Stores*

There are four basic characteristics of return logistics. These are (Wang et al. 2013):

1. Uncertainty—it is quite difficult to predict and calculate the routine due to the uncertainty of the amount, type, time of origin and distribution of reverse logistics.
2. High process costs—refunds and exchanges are random and therefore it is difficult to allocate storage and transport proportionately, precisely because of the random transfer of products.
3. Complexity—the exchange and return process is accompanied by complex procedures with a complicated structure and changing form, and each stage is influenced by the results of return logistics planning.
4. Low speed—the whole process is slow due to the complexity of the processing process. The process includes testing, judgment, and classification and is a demanding and time-consuming process.

The key reasons for the need for a return logistics process are (1) Information asymmetry: in electronic commerce, it is difficult for customers to get a complete insight into the characteristics of the purchased goods, since the perception is based on images seen on the Internet. The case of difference between the actual product and what is represented in the picture is one of the main causes of product returns; (2) Competition between online retailers: to present themselves better than the competition and thus attract as many customers as possible, some online retailers emphasize the option of a possible return in case of dissatisfaction, which attracts many customers, but on the other hand results in various return problems; (3) Goods problems—various conditions affect return problems, such as the deviation of the item from its description, defects or quality problems, inappropriate size, quantity, or even the item, goods that are close to or exceed the expiration date, damage caused by careless distribution; (4) Unstable consumer preferences—buying out of curiosity or impulse is prone to returns; and (5) Imperfections in logistics—certain occurrences of damage to goods may occur in the process of transportation, loading and unloading, handling and distribution. Components may be lost, goods may expire or spoil if not kept in the proper conditions.

The challenges that lie in the management of return logistics are (1) Online retailers do not attach enough importance to return logistics: in the traditional way of doing business, most retailers' main goal is to attract as many customers as possible.

In the case of return logistics, most companies are confused about the process, and the large number of returns, different items and locations only increases the difficulty in managing them; (2) Lack of an effective returns mechanism—customers are often suspicious of how returns are handled, so a company's return performance has a big impact on a customer's decision whether to make a purchase on a particular website; (3) Low processing efficiency and high costs: many companies have to build their own expensive return logistics management systems. Excessive costs of return logistics prevent companies from being enthusiastic about managing return logistics, which results in a decrease in the number of satisfied customers and ultimately lower profits; and (4) Poor quality of return logistics service: in some cases, it is difficult for the customer to get in touch with the customer service staff, and even when this happens, it can still be difficult to get a satisfactory result of the exchange processing (Wang et al. 2013).

3 Research

For the successful operation of online stores, apart from the high-quality, diverse and affordable range of products they offer, the satisfaction of their customers is also crucial. In the age of the developed system of online shopping, when customers are increasingly choosing this way of purchasing because it is increasingly available to them, the process of returning products is also inevitable. This process should run smoothly so that it does not make it difficult for companies to do business and so that precisely because of problems with returning products, customers do not give up this way of shopping, which every company wants to avoid.

There are many reasons why a customer can request a return: because the product does not match the description on the webshop, he does not like it for any reason, it does not suit him, the quality of the product is bad or he simply changes his mind about the ordered product. In addition, the product can be damaged during delivery, in which case there is again a necessary return process, as well as a challenge for stores on how to solve such a problem. For this very reason, a company engaged in online trade must have a well-organized product return system (https://europa.eu/youreurope/citizens/consumers/shopping/guarantees-returns/index_hr.htm).

In this paper, two studies were conducted. One is intended for companies that use the option of selling products online in their business, and the other is intended for customers who order products through online stores. The research aims to find out how companies handle the return process, what is most often returned and for what reason, their satisfaction with the process itself and their readiness for improvement as well as ways to improve their own business. In parallel with that survey, a user survey was also conducted to better analyze the differences in opinion and experience of users and companies. Also, a second survey is needed to examine the attitudes of customers regarding their product return habits.

3.1 Methodology and Data Collection

For the purpose of this research, data was collected using a questionnaire created through the Google Forms (https://www.google.com). The first survey intended for companies in the field of web commerce had 67 participants, and the second, intended for users of online stores, had 68 respondents. The survey for companies was sent to 480 e-mail addresses of companies in the field of web trade that were found through the Digital Chamber (Digitalna komora 2024) and through a website with a printout of online stores in Croatia, which would mean that 13.96% of the companies that received the survey and answered the same. The survey for users was sent to 450 e-mail addresses, so 68 answers make up a response of 15.11%. The first survey consisted of 31 questions, and the second 27, and it took an average of 10 min to solve them. Both surveys were anonymous. The surveys mainly consisted of simple questions with already provided answers, from which we find out which branches of sales the companies deal with, what are the main reasons for returns in a certain company, the way products are managed, the time required to execute returns, the most important aspects in the logistics of delivery and returns, as well as ways to track packages and communicate with customers. In the first survey, statements were used that the companies had to evaluate on a Likert scale from 1 to 5, and from these questions data were obtained on the satisfaction of the companies in cooperation with their logistics partners, the efficiency and ability of the package tracking method, the assessment of the satisfaction of their customers as well as workers in the webshop and the willingness to invest in the improvement of the processes themselves, that is, the use of new technologies to improve the process. Also, several short yes/no questions were asked to find out if the companies have an established return policy and if they are planning improvements in the future.

The second survey contains the three types of questions already mentioned, however, this survey also includes a question to which the respondent answers himself, which is a question about the age of the respondent. The questions with already set answers in this survey give us information about the respondent's gender, the frequency of online shopping, the number of returns in the last year, as well as the most common products that are returned and for what reason, the time required to execute the return, as well as the reasons why they decide to do this way of shopping. The third type of question found in the second survey are statements that the respondents should mark on a Likert scale from 1 to 5. The goal of these questions is to find out how satisfied customers are in general with the speed of returns, the return process itself, and communication with the web store. From the same type of question, information was also obtained about how much importance they attach to the fact that the return is free of charge, that it is fast, environmentally friendly, simple and that all conditions of the return are clearly indicated. In the same way, their familiarity with the use of new technologies in the process of delivery and returns is also learned, as well as the opinion that such technologies would improve the process itself. The fourth type of questions in this survey were yes/no questions about whether customers contacted customer service during the return process, whether they successfully returned the product, whether they were aware of their rights, and whether they had ever experienced delivery or returns using new technologies, i.e. drones and mobile robots.

4 Results

The survey, intended for companies, was completed by stores that sell various products. Among the most frequent participants were companies that offer clothing and fashion products, and there were 14 such companies, followed by stores that sell technical equipment such as computers and mobile devices, 11 of them, followed by 10 companies from the field of cosmetics, 8 of those that deal in the sale of household appliances as well as 8 who sell medicines and nutritional supplements, 7 stores deal in the sale of food and food products, 4 companies sell books and 2 stores sell food and equipment for pets. The rest of the answers were from companies that sell other products such as toys, fitness equipment, baby equipment, mattresses, car equipment, party programs, musical instruments (Fig. 1) It was possible to choose several answers to the first question and for this reason, the number of answers is greater than the number of respondents. It is evident from the answers that the survey covers a large part of different web shops, which gives a broader picture of the overall analysis.

4.1 Survey Results for Companies

When asked which product category is most often returned, the most common answer was the following: clothing and fashion products, followed by technical equipment, cosmetics and household appliances. Since most of the companies engaged in online sales belonged to these categories, such a number of responses was expected.

When asked to state the main reason for returns in a certain store (Fig. 2), companies stated that the most common reason is that customers return a product because they have changed their mind about it. In addition, common reasons for returns are that the product does not work properly or was damaged during delivery. Also, one of the frequent reasons stated in the questionnaire is that the customer does not like the size of the ordered product, so he has to return the product. A small part of the companies answered this question and that the most common reason for returns is precisely that customers give up the purchase, i.e. use their right to order the product, see it live and return it without any financial consequences and without having to leave the house to do so.

Of the offered answers that online store return methods enable their customers, the money refund option received the most answers, as many as 64. The product exchange option received 57 answers, while the purchase voucher option was the

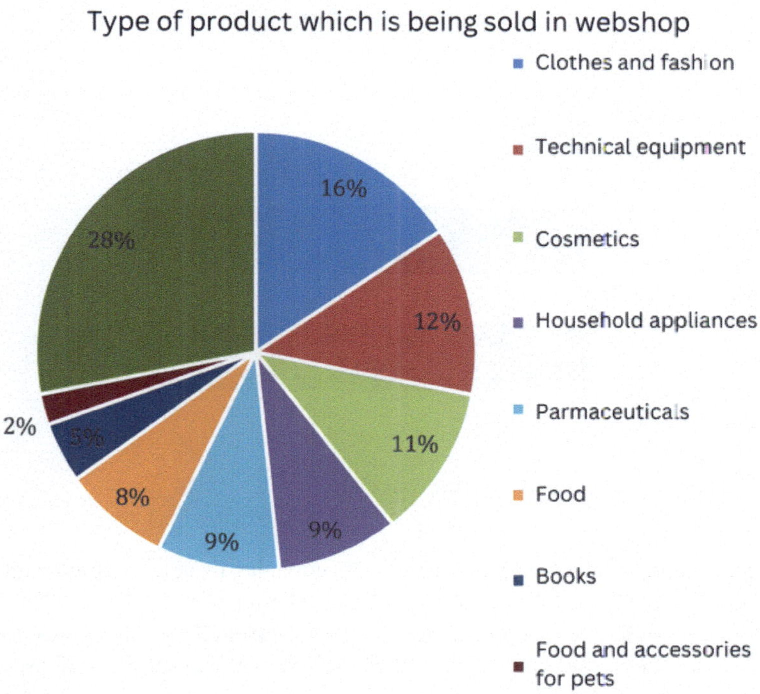

Fig. 1 Type of product which is being sold in the webshop of the companies, own study

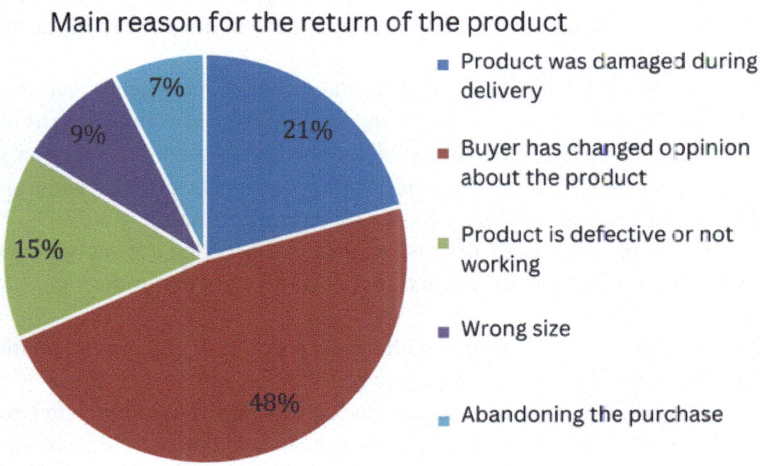

Fig. 2 Main reasons for the return of the product, own study

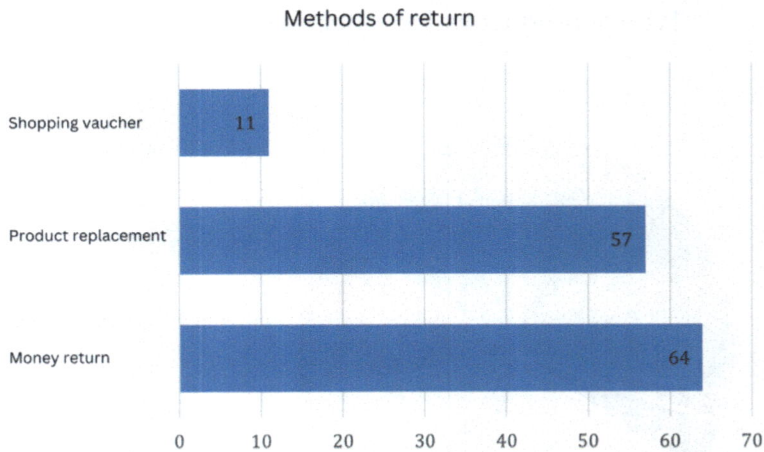

Fig. 3 Main methods of the product return, own study

least represented, with 11 answers (Fig. 3). Companies usually communicate with customers during the return process via e-mail and phone.

The time it takes for a company to resolve a return from the moment the customer reports it until the refund or replacement of the product is usually 3–6 days, and as many as 38 companies chose that answer. The second most common answer was that it takes less than 3 days to return or exchange a product, and 19 companies agree with this answer, while 9 companies claim that the process takes 1–2 weeks. Only one company responded that it takes more than 2 weeks for the refund or product replacement process.

In most stores, there is a certain time frame in which a customer can request a return. According to the survey, in most companies this period of time is up to 14 days from when the customer receives the order, the second most common answer is that the company does not have a specific deadline for returning the product, while another 10% of companies have a deadline of 30 or 60 days in 2% online store.

To manage the product return process, companies most often use internal logistics, and only one company that participated in the survey said that it does not have an organized return process.

Companies are generally satisfied with the current ability to monitor and manage products and the average satisfaction score is 3.91 (Fig. 4).

However, regardless of satisfaction with the current ability to monitor and manage processes, as many as 45% of companies have plans and desire to improve them in the future.

The biggest challenge in cooperation with logistics partners, and stores that deal with online sales, is the delay in delivery. It is very important to inform the customer when the shipment will arrive, however, due to the increasing number of such orders, the delivery services are often overloaded, so the deadlines are extended. For this reason, a company must have a reliable way to track the packages they send to

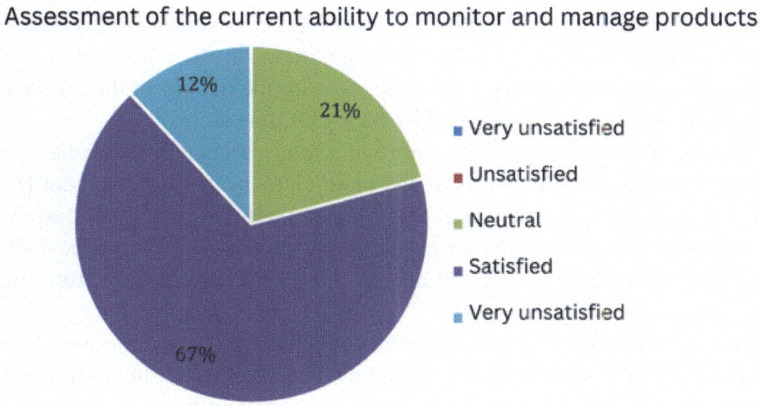

Fig. 4 Assessment of the current ability to monitor and manage products, own study

customers, as well as to track those packages that customers return. In this way, you can always know where the package is and packages cannot be lost. Regardless of the importance of the above, there are still certain companies that do not track the status of shipments. In this research, 3% are such, while not a single company uses specialized software solutions for tracking the status and location of packages (Fig. 5).

When returning the product, it is possible that the product is damaged in the process itself. When asked about the frequency of such a situation in their company, the companies answered that fortunately such situations are rare. If there is damage

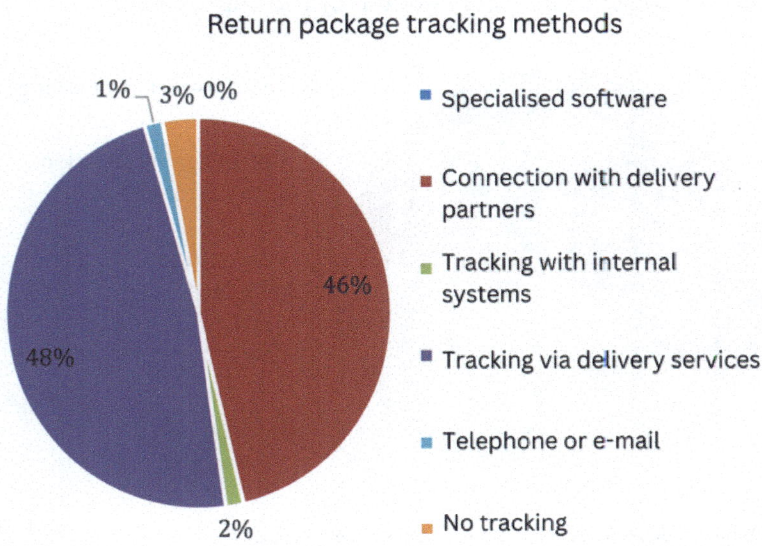

Fig. 5 Return package tracking, own study

during the delivery of the product to the customer, companies usually replace the product or refund the money to the customer. Also, in the surveyed companies, product returns due to customer dissatisfaction with the quality of the product itself happen rarely, and this was confirmed by 64% of online stores.

As the most important aspects for improving return logistics, companies state the reduction of the number of returns as the main element for improvement (Fig. 6). The second most common answer is to reduce the number of returns, followed by better communication with customers and faster processing of returns. Only two companies answered that they do not plan or think about improving return logistics at all.

In modern times, new delivery and return methods are being imposed, however, most companies are not yet well acquainted with such digital technologies and most of the respondents marked the statement that they are familiar with the use of new digital technologies in supply and return logistics as if they did not even agree with it, nor do they agree, but that they are neutral. Also, companies are mostly neutral regarding the claim that mobile robots and drones would improve the delivery and return experience. According to the attitudes of the companies, it can actually be seen that it leans more in the negative direction, since only 9 companies gave a positive answer regarding the use of new digital technologies—drones and mobile robots, that is, how they agree that their use would improve the delivery and return process, while 52% of companies said they were neutral.

The majority of companies, as many as 44 of them, believe that the workers in their web shop are satisfied, and the average score with which the companies evaluated the satisfaction of their workers is 3.97. Companies point to prolonged sitting as the

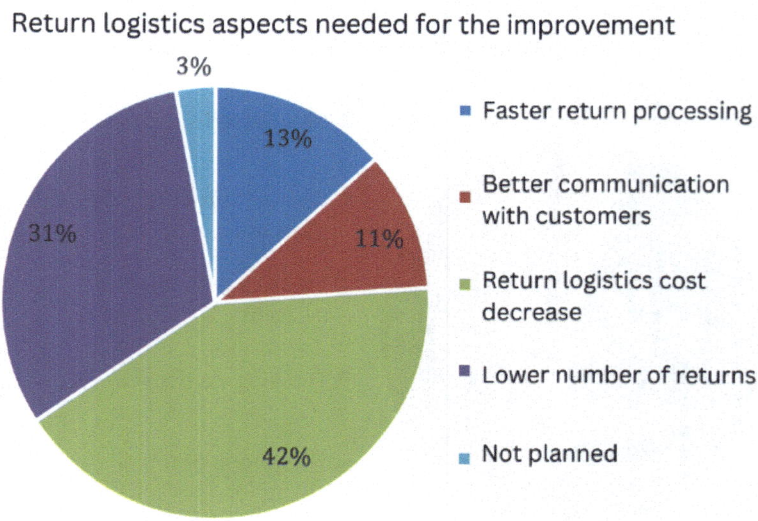

Fig. 6 Return logistics aspects needed for improvement, own study

biggest problem workers face, however, 46% of companies say they make a big effort to improve the working conditions of their employees.

4.2 Results of the Customer Survey

In parallel with the survey intended for companies engaged in online trade, a second survey was created and sent to customers who use e-commerce services. 68 respondents between the ages of 19 and 69 took part in that survey, and most of them were between the ages of 22 and 27, 39 of them, which makes up 57% of the respondents. 41 women took part in the examination, which is 60% of the respondents and 27 men, that is 40%.

When asked how often they order products through web shops, respondents in 25% of cases said that they do it sometimes, 21 people said that they do it often, and 18 people said that they do it very often. Only 6% of people answered that they rarely order products through online stores.

All respondents encountered the situation that the ordered product did not suit them for some reason and therefore decided to return the product. More than half of the respondents, 56% of them, returned at least one product in the last 12 months. Then follow the answers of those who returned the products twice, and there were 9 of them, which is 13% of the respondents, there were 12 users with 3 and 4 returns in the last year, that is, 6 respondents with 3 and 6 respondents with 4 returns. The next group of respondents with 5 or more returns in the last year includes 9 people, making up 13% of the responses. Respondents mostly answered that they order clothes and fashion products through online stores, so it is not surprising that the most returns are products from that category, as much as 72%. This is followed by the category of technical equipment with 16%, household appliances with 3% and other with 9%, which includes cosmetics, medicines and food supplements and tools.

The reason why customers mostly decide to return is that they changed their mind about what they ordered, and this was the answer given by 29 respondents. In addition, customers also cite poor product quality (17 respondents), damaged products (10 respondents), delivery delay (5 respondents), inappropriate size (11 respondents) and that the product does not match the description on the web shop (16 respondents) as reasons for return (Fig. 7). To the same question, the companies also answered that it is precisely the customer's change of mind that is the reason for returning the product.

Unlike companies, which mostly claim that it takes them 3–6 days to process a return from the moment the customer reports it to the refund or replacement of the product, users say that in practice it is not quite like that. As many as 40% of customers claim that this process takes 1–2 weeks, while only 28% of respondents agreed with the most common answer from the company, that it takes 3–6 days (Fig. 8). Customer experiences show that situations where returns are made in less than 3 days are extremely rare, and only 3% of respondents answered this way. Also,

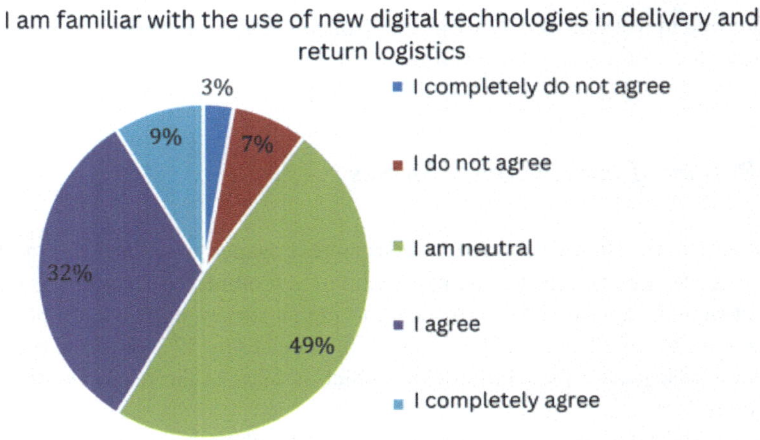

Fig. 7 Familiarity of novel technologies in delivery and return logistics, own study

19% of respondents encountered situations in which the refund takes more than 2 weeks (Figs. 9 and 10).

One of the main features of online shopping is its practicality, i.e. that shopping can be done very easily and quickly from the comfort of your own home at any time of the day or night. Therefore, it is not surprising that respondents chose practicality as the main motivator for online shopping, and even 48 respondents chose that answer. In addition, online stores offer a wide selection of products, which 40 respondents agree with. The next most important motivator is a more favorable price, since online stores do not have to have as many employees as is the case in physical stores and do not pay for the rent of space for the store, they can offer much more favorable prices than is the case in physical stores, and this is exactly the answer that's why

Fig. 8 Reasons for the return, own study

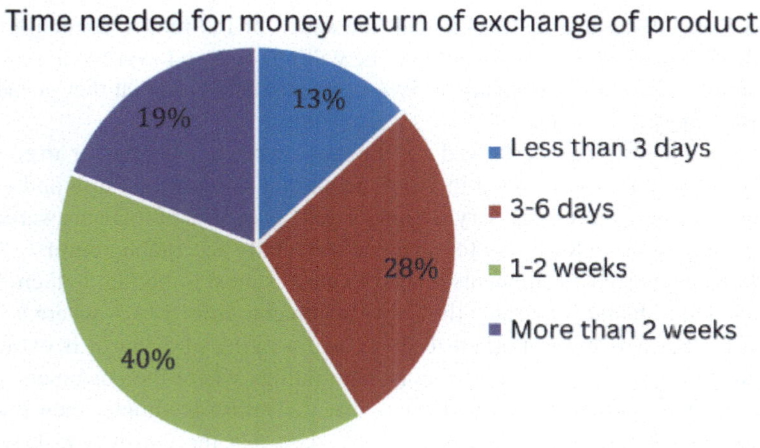

Fig. 9 Time needed for the money return of product exchange, own study

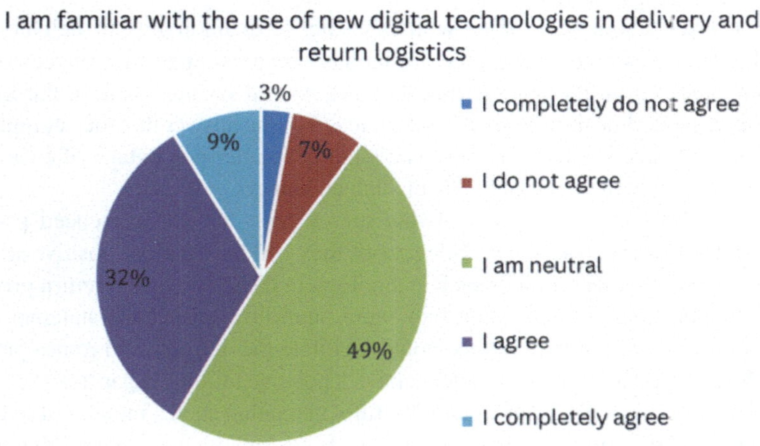

Fig. 10 Customer familiarity with use of novel digital technologies in delivery and return logistics, own study

34 respondents marked it as a motivator for online shopping. Then according to the number of answers, there are fast delivery and reviews from other customers, each with 11 answers, and only one answer is due to previous positive experiences with a particular web store.

When asked about satisfaction with certain aspects of the return, respondents were able to rate their satisfaction with scores from 1 to 5, where a score of 1 represented that the customer was very dissatisfied, and a score of 5 was very satisfied. The average score given by 68 respondents regarding communication with the web store is 3.82. The overall product return process was rated the best with a rating of 3.96, while the speed of the return itself received the lowest average rating of 3.71.

Regardless of the dissatisfaction with the speed of the return, the majority of respondents answered that they would not be willing to pay a lower fee to make the return simpler and faster. As many as 59% of respondents said that they would not be willing to pay a lower fee.

However, the respondents showed a willingness to pay a lower fee for an environmentally friendly return, keeping in mind the current state of the Earth and understanding that changes are necessary. 38 respondents, i.e. 56% of them answered that they are ready to pay a lower fee for an environmentally acceptable return.

Users then received 5 statements related to what is most important to them when returning. They should have rated the above statements from 1 to 5, where a rating of 1 means that it is not important to them, and a rating of 5 that it is extremely important to them. Table 1 shows the average ratings where the customers stated that the most important thing for them is that the return is simple, then that the conditions of the return are clearly indicated, then that the return is fast, so that there is no additional charge. The statement that it is important for the return to be environmentally acceptable was rated the worst by customers and this statement received an average rating of 3.27.

When asked about familiarity with new digital technologies in delivery and return logistics, customers gave very similar answers to companies. Both customers and companies chose the answer that they are neutral on this issue in the largest percentage, and unlike the companies who had 8% of the answer that they completely agree with the stated statement, there is a slightly smaller percentage of customers who claim this, i.e. there are only 3% of such customers.

Although 99% of respondents of this survey have never experienced product delivery or return via drones or mobile robots, they still have a more positive attitude regarding the implementation of such technology in the delivery and return process. 45% of respondents answered that they were neutral regarding the statement that the use of drones and mobile robots would facilitate the delivery and return process itself. However, 35% of the respondents have a positive attitude regarding the stated statement and 26% of them agree, and 9% fully agree that such a method of delivery and return would improve logistics processes. If customers were offered the option of delivery via drones and mobile robots, 65% of them would be ready to try it.

Table 1 Average grades of the customer assessment of importance of the several aspects of return, own study

When returning, it is important to me that there is no additional charge	4,35
When returning, it is important for me that it is environmentally friendly	3,27
When it comes to returns, it is important to me that it is fast	4,38
When it comes to returns, it's important to me that it's simple	4,76
When returning, it is important to me that the conditions are clearly indicated	4,65

5 Conclusion

Returns logistics in e-commerce can be considered a key aspect of the entire supply and returns chain and is of utmost importance as it directly affects customer satisfaction. This process must be efficient, because in this way, consumer trust is achieved and a positive image of a certain web store is maintained. Speed, simplicity and transparency are key factors influencing the customer experience in the product return process.

To optimize return logistics, it is necessary to integrate an approach that includes technological solutions as well as to build an exceptionally good communication channel between customers and stores. Another key factor for the success of a web store is cooperation with reliable logistics partners. When a company cooperates with such partners, there is the possibility of tracking shipments and clear guidelines for returns, difficulties are significantly reduced, and then both customers and retailers themselves are more satisfied.

Investing in the improvement of return logistics will not only contribute to customer satisfaction, but ultimately also bring long-term profit and competitiveness of the company on the market.

With the help of surveys, the attitudes and habits of customers who returned a product at least once, as well as companies that sell products via the Internet, were examined. Although the survey combined companies from different sales sectors, as well as customers with different buying habits and different ages, their answers mostly do not differ. Respondents to both surveys confirmed that speed and ease of return are the aspects they value the most.

Regardless of overall customer satisfaction, there are still many aspects that can be improved to make returns even simpler, faster and ultimately cheaper. For the future research, the provided results can be used as a framework to improve optimize logistics processes of certain companies and businesses by the Industry 5.0 standards to improve overall efficiency and both worker and customer satisfaction level.

References

Aćimović S, Mijuskovic VM, Vuksanović Herceg I (2020) Factor determination and reverse logistics modelling: Theory vs practice

Caniëls MCJ, Gehrsitz MH, Semeijn J (2013) Participation of suppliers in greening supply chains: An empirical analysis of German automotive suppliers. J Purch Supply Manag 19(3)

Chavez R, Yu W, Feng M, Wiengarten F (2016) The effect of customer-centric green supply chain management on operational performance and customer satisfaction. Bus Strategy Environ 25(3)

Dangelico RM, Pontrandolfo P (2015) Being green and competitive: The impact of environmental actions and collaborations on firm performance. Bus Strategy Environ 24(6)

Digitalna komora. https://digitalnakomora.hr/ (6 May 2024)

Gimeney C, Sierra V (2013) Sustainable supply chains: Governance mechanisms to greening suppliers. J Bus Ethics 116(1)

https://europa.eu/youreurope/citizens/consumers/shopping/guarantees-returns/index_hr.htm. (3 May 2023)

https://www.google.com/forms/about/. (5 May 2024)

Ivanova T, Rogaczewski R, Lutsenko I (2022) Influence of reverse logistics on competitiveness, economic performance, ecological environment and society. Logforum 18(1):5

Lambert D, Stock R (1981) Strategic Physical Distribution Management, Homewood. Irwin, USA

Murphy P, Poist R (1989) Management of logistical Retro-movements: An empirical analysis of literature suggestions, transportation research forum

Patak M, Branska L, Pecinova Z (2020) Importance of reverse logistics services as an antecedent for building a green supply chain, Pardubice

Petljak K, Zulaus K, Štulec I, Seuring S, Wagner R (2018) Green supply chain management in food and retailing: Survey-based evidence in Croatia

Stock JR (1992) Reverse logistics, Oak Brook: Council of logistics management

Wang W, Liu Y, Wei Y (2013) Research on management strategies of reverse logistics in e-commerce environments. J Syst Manag Sci, Qingdao

Waters D (2003) Logistics: An Introduction to Supply Chain Management. Palgrave Macmillan, Houndmills

Yen YX (2018) Buyer-supplier collaboration in green practices: The driving effects from stakeholders. Bus Strategy Environ 27(8)

Building Sustainably: Understanding Green Supply Chain Management in Angola's Construction Small Enterprises Through Grounded Theory

Chitandula Lutock Liahuka and Noémi Piricz

Abstract The urgent need to address environmental issues has become a focal point in global discussions. There is a growing interest in finding innovative strategies to reduce environmental degradation, especially in highly polluting industries like construction. This study focused on the adoption of green supply chain management practices among small construction companies in Angola, using a grounded theory approach. We conducted in-depth interviews with professionals from 12 small firms to understand their experiences, perceptions, and strategies related to sustainable practices. The results reveal several barriers to implementing sustainable practices, such as financial constraints, lack of knowledge, and economic instability. Other challenges include limited finances, the high costs of green technologies, and insufficient training and awareness about GSCM. In addition, economic factors like inflation worsened these issues, making it hard to prioritize long-term sustainability investments. The study also highlighted the crucial role of governmental support through regulatory frameworks, financial incentives, and training programs in facilitating GSCM adoption. The benefits of adopting GSCM practices were evident, including long-term cost savings, improved efficiency, and natural resource conservation, which provide strong incentives for adoption. The resulting grounded theory model suggests that a comprehensive approach incorporating targeted financial incentives, capacity-building initiatives, and robust regulations could effectively encourage the adoption of sustainable practices. This framework has significant implications for promoting environmentally responsible construction in Angola and other developing nations dealing with sustainable industrialization. The study addresses a critical research gap by focusing on small enterprises in an underexplored context and provides valuable insights that can inform future research and practical interventions in similar settings.

C. L. Liahuka
Safety and Security Science, Óbuda University, Budapest, Hungary
e-mail: liahuka@stud.uni-obuda.hu

N. Piricz (✉)
Keleti Károly Faculty of Business and Management, Óbuda University, Budapest, Hungary
e-mail: piricz.noemi@kgk.uni-obuda.hu

A. Kolinski and M. Adamczak (eds.), *Digitalisation of the Greening Supply Chain*, EcoProduction, https://doi.org/10.1007/978-3-031-88918-9_4

Keywords Green supply chain management · Construction small enterprises · Angola · Grounded theory

1 Introduction

In recent decades, the relentless pursuit of urbanization and infrastructure development has led to an unprecedented expansion of the global construction industry, contributing an estimated \$10 trillion to the world economy annually and 13% of GDP contribution (Barbosa et al. 2017). However, this exponential growth has come at a concerning environmental cost, with the sector now accounting for 39% of global carbon emissions, 40% of the world's energy consumption, 25% of the world's water consumption, 40% of the global waste generation, and substantial depletion of natural resources (International Energy Agency 2019; UNEP 2016). These environmental issues that have greatly affected the supply chain do not exclusively impact the construction industry, as other industries worldwide face the same problem.

As more people become aware of environmental issues, customers are demanding that companies act more responsibly by adopting sustainable practices to protect the environment (Mutingi et al. 2014; Sarkis et al. 2011). This shift in consumer expectations has given rise to green supply chain management (GSCM), a concept designed to tackle the environmental problems that traditional supply chains often overlook.

Green supply chain management incorporates eco-friendly practices throughout the supply chain (Fahimnia et al. 2015; Srivastava 2007). This comprehensive approach not only aims to reduce the environmental impact but also promotes sustainability from start to finish (Hervani et al. 2005). Implementing these green practices can benefit organizations by cutting costs, enhancing their reputation with stakeholders, improving operational efficiency, and reducing waste (Agyepong and Nhamo 2017; Benmamoun et al. 2017; Hervani et al. 2005).

In the construction industry, adopting green practices is crucial for reducing environmental impact and fostering sustainable development (Mingqiang and Zuxu 2011). In Angola, where the construction sector has been one of the most important factors in the country's "rebirth" and economic resurgence following the more than 25-year-long civil war, the environmental impact of this sector is a growing concern (James 2020; Pautasso 2010). Although green supply chain management is still in its early stages in Angola, it is essential for addressing issues like deforestation, air pollution, and unsustainable resource use, which have been exacerbated by the rapid post-war reconstruction boom and substantial infrastructure investments (African Development Bank 2018; UNEP 2021).

In the construction industry, adopting green practices is crucial for reducing environmental impact and fostering sustainable development (Mingqiang and Zuxu 2011). In Angola, where the construction sector has been one of the most important factors in the country's "rebirth" and economic resurgence following the more than 25-year-long civil war, the environmental impact of this sector is a growing concern

(James 2020; Pautasso 2010). Although green supply chain management is still in its early stages in Angola, it is essential for addressing issues like deforestation, air pollution, and unsustainable resource use, which have been exacerbated by the rapid post-war reconstruction boom and substantial infrastructure investments (African Development Bank 2018; UNEP 2021).

Despite its potential, the adoption of green supply chain management practices in Angola's small construction enterprises remains limited. Challenges such as low awareness, scarce resources, and regulatory barriers hinder progress, similar to other emerging economies (Tembo-Mwanaumo et al. 2023; Niemann et al. 2016). Most existing research on GSCM in Africa focuses on large corporations, leaving a gap in understanding its application in small enterprises. This gap highlights the need for more comprehensive research linking environmental sustainability with construction activities in developing countries like Angola.

Therefore, this qualitative grounded theory research aims to fill a gap in the literature by developing a theoretical model based on qualitative data to reveal the challenges faced by small construction enterprises in Angola in adopting green supply chain management practices. The primary objective was to gain a deep understanding of how these companies perceive and are aware of GSCM, the challenges they face, and how they manage the implementation of these practices in their daily operations.

The study aimed to answer the following research questions:

1. What is the level of awareness and understanding of Green Supply Chain Management (GSCM) concepts within these small firms?
2. What are the main barriers and challenges faced by small construction enterprises in Angola in adopting GSCM practices?
3. What are the perceived benefits of adopting GSCM practices for small construction enterprises in Angola?

A grounded theory approach was chosen because it allows for the generation of a theory grounded in the data itself and is particularly suitable for exploratory research in areas with limited or insufficient existing theory (Glaser and Strauss 1967; Corbin and Strauss 2008). Recognizing the significant role of the construction industry in Angola's economy and its substantial environmental impact, the research aimed to contribute to the sustainable development discussion by focusing on the following objectives:

I. To explore the challenges of implementing GSCM practices among owners and managers of small construction enterprises in Angola.
II. To assess the level of awareness and understanding of GSCM concepts within these small firms, establish a baseline of any existing sustainable practices, and identify key environmental concerns relevant to their operations.
III. To uncover potential benefits and support systems available for small construction companies considering the adoption of GSCM practices.

This study makes significant contributions to the body of knowledge in several ways. Firstly, by exploring the adoption of GSCM practices in small construction businesses in Angola, it addresses a critical research gap in an underexplored context, providing valuable insights into the unique challenges and dynamics faced by these firms. Secondly, the development of a grounded theory model offers a theoretical foundation for understanding the factors influencing GSCM implementation in similar contexts, informing future research and practical interventions.

2 Literautre Review

2.1 Green Supply Chain Management (GSCM)

Green Supply Chain Management (GSCM) was introduced as an innovative approach to integrating environmentally sound practices into every phase of the supply chain, from product design to end-of-life management. It aligns organizational environmental objectives with stakeholder demands for sustainable economic, social, and environmental practices (Fahimnia et al. 2015; Hervani et al. 2005; Srivastava 2007; Zhu et al. 2007, Sosnowski 2022). Implementing GSCM practices helps organizations reduce carbon emissions, transition towards a circular economy through increased use of recycled and reused materials, and eliminate hazardous substances from products and processes, thereby protecting the environment and public health (Mafini and Muposhi 2017; Muvaka and Henry 2018; Sarkis 2003).

The construction industry significantly contributes to environmental degradation through substantial greenhouse gas emissions, resource depletion, and waste generation (International Energy Agency 2019; UNEP 2016). These issues underscore the need for sustainable practices like green supply chain management, which aim to minimize the environmental impact of supply chain operations (Fahimnia et al. 2015; Srivastava 2007).

The integration of sustainable materials and practices in construction projects, such as the use of alkali-activated materials, sustainable construction materials like Compacted Mineral Mixtures (CMMs), and modular construction systems, can significantly mitigate environmental damage and improve resource efficiency (Ali et al. 2023; Bühler et al. 2023; Dhawan et al. 2023; Jeong 2023).

2.2 Challenges and Opportunities in the Construction Industry

The construction industry is vital to the economic development of any country due to its significant impact on GDP, employment generation, and infrastructure development (Turin 1973). Post-war Angola experienced rapid economic growth, particularly between 2002 and 2014, driven by substantial investments in infrastructure reconstruction. The construction sector contributes approximately 7% to Angola's GDP, reflecting its crucial role in economic modernization (African Development Bank

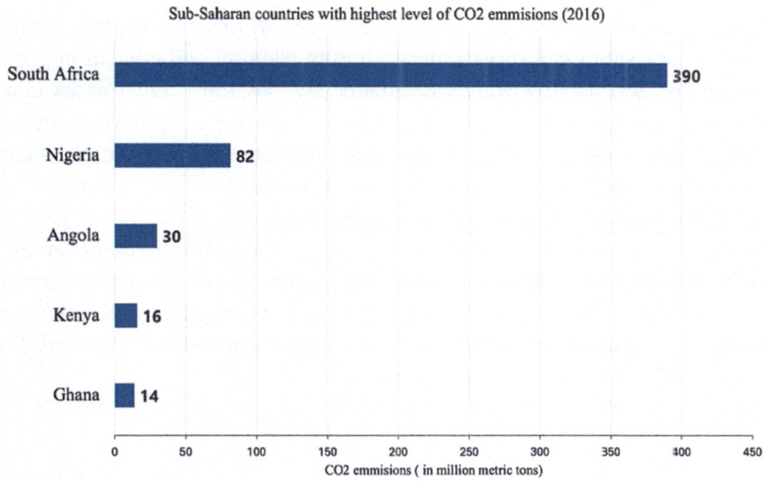

Fig. 1 Emission of CO_2 by countries in the Sub-Saharan region, based on: (Worldometer 2023)

2020). However, despite its economic importance, Angola's construction industry faces significant challenges.

Regulatory barriers, bureaucratic red tape, and inconsistent enforcement lead to delays and increased costs for construction companies (Del Passo 2019; Søreide 2011). The sharp decline in oil prices from 2014 to 2016 further strained the sector, resulting in budget cuts and project delays (Wanda et al. 2023). Additionally, a shortage of skilled labor and inadequate workforce training hinder productivity and quality in construction projects. Improving vocational training and education programs is essential to develop a more skilled workforce capable of meeting industry demands (Ra et al. 2015; Wanda et al. 2023).

The National Development Plan (PND) 2018–2022 outlines priorities such as road rehabilitation and new housing to accommodate urbanization (Government of Angola 2018). The China-Angola investment model, where China finances infrastructure in exchange for natural resources, is another example of Angola's infrastructure ambitions (Begu et al. 2018; Ye and Ye 2013). However, this rapid growth has resulted in significant environmental challenges, including deforestation, soil erosion, and increased greenhouse gas emissions, making Angola one of the biggest emitters of CO_2 in the Sub-Saharan region, as shown in Fig. 1 (UNEP 2013; Worldometer 2023).

2.3 *Small Enterprises and GSCM Adoption*

Small enterprises worldwide face unique challenges in adopting GSCM practices, including resource constraints, financial limitations, lack of expertise, and limited access to green technologies (Azevedo et al. 2011; Walker et al. 2008). There is often a lack of awareness and understanding of GSCM principles among managers and employees of small enterprises, necessitating training and education programs (Simpson et al. 2007). However, the smaller size of these enterprises can also present advantages, such as greater flexibility and quicker adaptation to new practices (Aragón-Correa et al. 2008). Many small construction firms source materials locally, reducing transportation emissions and supporting the local economy, which aligns well with GSCM principles (Rao and Holt 2005).

In Angola, the adoption of GSCM practices among small construction enterprises is still in its infancy. Limited awareness, scarce resources, and regulatory barriers are significant obstacles. The integration of green practices is crucial for these enterprises to mitigate environmental impacts and enhance their sustainability (Ethel Tembo-Mwanaumo et al. 2023; Niemann et al. 2016).

3 Methods

A research design serves as a roadmap guiding a researcher from the beginning to the end of a study (Yin 2009). To comprehensively understand the adoption of green supply chain management (GSCM) in Angola's construction industry, particularly among small companies, this study adopts a qualitative research methodology based on the grounded theory approach. This systematic approach generates theory from data, making it suitable for exploratory research in areas where existing theory is lacking or insufficient (Dunne 2011; Glaser and Strauss 2017).

According to Bitsch (2005), the grounded theory development process involves several stages, such as identifying a research problem, formulating the research question, gathering data, coding and analyzing the data, and finally, constructing the theory.

The diagram in Fig. 2 shows the research methodology used in this study. First, the research problem is defined, followed by defining the research questions. This is then followed by data collection and theoretical sampling. Grounded theory is employed to systematically analyze in-depth interview materials. The data analysis process involves several steps, including open coding, axial coding, and selective coding. If theoretical saturation is not achieved, more data is collected. Once saturation is reached.

3.1 Population and Sample

In this study, we used snowball sampling as our main method for recruiting participants. This method is effective for reaching populations that are hard to access through traditional methods (Noy 2008). The construction industry in Angola is made up of formal and informal networks, with many small businesses operating in close communities where traditional sampling methods may not work well. We used this

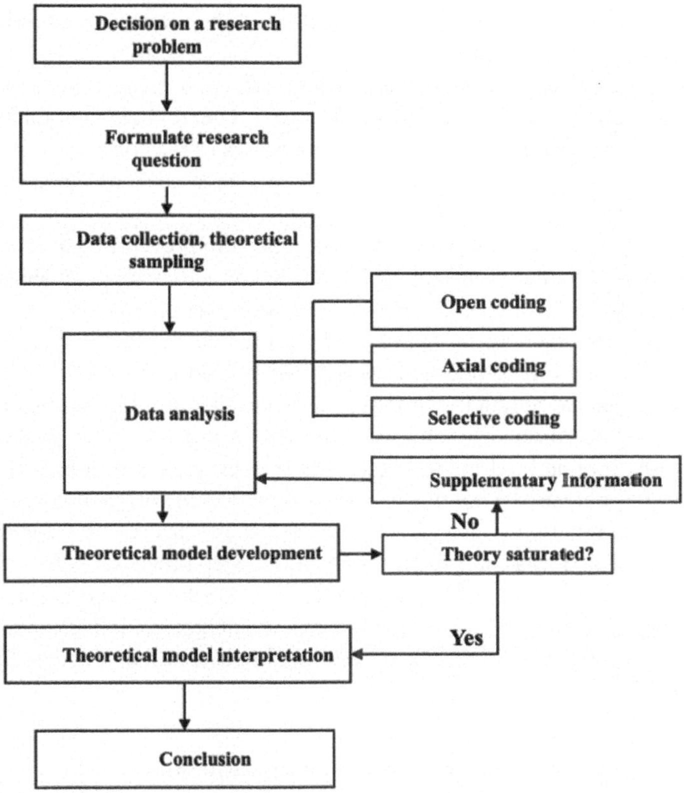

Fig. 2 Research Flowchart for Grounded Theory Approach, based on (Adapted from Bitsch 2005)

approach to make use of personal and professional connections to find participants with valuable industry knowledge who could speak and understand Portuguese or English. Initially, we contacted experienced individuals with a basic understanding of green supply chain management (GSCM) through industry associations, social media, and personal contacts. After getting their consent, these participants recommended others in their networks, resulting in a diverse and well-informed sample (Atkinson and Flint 2001).

3.2 Data Collection

From May 2023 to April 2024, primary data for this research were gathered through in-depth interviews with owners and managers of 12 different small construction companies operating across Angola, specifically in Luanda, Benguela, and Huambo. However, due to communication challenges, the majority of interviews were conducted via telephone and Zoom calls, and the remaining interviews were conducted face-to-face in April 2024.

To ensure the confidentiality of the participants, each company was assigned a unique identifier (SC1 to SC12) rather than using their real names (Table 1). This anonymization was crucial in protecting the identities of the companies and encouraging open and honest discussions about their experiences and challenges with green supply chain management.

During every interview, diligent notes were taken to prevent the loss of any critical information. The main focus of each interview was to gain an understanding of the challenges faced by small construction companies in implementing green supply chain management practices in Angola and it's the benefits that come from it. All interviews were audio-recorded and transcribed verbatim for analysis. The interviews were originally conducted in Portuguese and then transcribed and translated into English. As mentioned before, the interviews were conducted in three ways: telephone calls, online, and face-to-face. These interviews time range from 35 to 50 min. For the analysis, 10 interviews were selected as the original coding analysis material, and the remaining interviews were used as the theoretical saturation.

The data collected was analyzed through a three-step process: open coding, axial coding, and selective coding. To maintain the trustworthiness and accuracy of the results, the study rigorously followed the steps of categorical induction and model construction during the coding process, resulting in the creation of codes, categories, and themes.

Table 1 Information related to the interviews and participants, own study

Sample group	Date	Company	Interviewee
First	May-November 2023	SC1	Owner
		SC2	Owner
		SC4	Manager
		SC5	Manager
Second	November (2023)-March (2024)	SC6	Owner
		SC7	Manager
		SC8	Manager
		SC9	Manager
		SC10	Manager
Third	April 2024	SC11	Owner
		SC12	Manager

4 Data Analysis

We conducted a general analytical procedure for a detailed analysis of the primary data. To ensure an accurate qualitative analysis of the interviews, we first transcribed them using Descript and then manually translated them into English in Microsoft Word. We paid special attention to accuracy during transcription and translation by conducting a comprehensive review of all the data. The transcripts were then categorized based on recurring themes and patterns. The coding process involved carefully reading the text, considering all its meanings, and creating categories, identifying interconnected themes, their meaning, and consequently interpreting them.

The data analysis process followed the three-stage coding process proposed by Corbin and Strauss (2008): open coding, axial coding, and selective coding.

In the open coding phase, the interview transcripts were analyzed line by line to identify concepts and categories related to the phenomenon, followed by axial coding, where these categories were refined and connected to form sub-themes. Finally, in the selective coding phase, a core theme that integrates the sub-themes and provides a comprehensive explanation of the phenomenon was developed. The data also underwent fragmentation, comparison, and categorization accordingly to develop themes, descriptions, and theories (Walker and Myrick 2006) related to the cyber supply chain.

4.1 Open Coding

Open coding is the initial stage of qualitative data analysis, where raw data from the interviews is broken down into discrete parts and examined closely. This process involves identifying and labeling concepts found in the data to create categories. For our study, we meticulously reviewed the interview transcripts, assigning labels to significant statements that highlighted various factors influencing sustainable practices in small companies. This stage laid the foundation by revealing common themes such as financial barriers, knowledge gaps, and economic challenges, as shown in Table 4 (Appendix A).

4.2 Axial Coding

Axial coding is the second stage where we reassemble the data by linking categories identified during open coding. This involves identifying relationships among the open codes to form more coherent and comprehensive categories. In our analysis, we connected related concepts to develop broader categories like financial barriers, knowledge gaps, and governmental support. With axial coding, we consolidated the

Table 2 Results of axial coding, own study

Main category	Category connotation
Financial barriers	High cost of green technologies, financial constraints, sustainability seen as luxury, limited financial resources
Knowledge gaps	Lack of training, limited awareness, focus on immediate concerns
Economic challenges	Economic instability, inflation, economic fluctuations
Governmental support	Regulatory support, incentives for training, government guidelines
Environmental and economic benefits	Conservation of natural resources, Long-term Cost Savings, improved efficiency

open codes into five main categories (Table 2), reflecting a structured understanding of the interactions and influences among different themes.

4.3 Selective Coding

Selective coding is the final stage, where we integrate and refine the categories developed during axial coding to form core categories. This involves identifying the central phenomenon and systematically relating it to other categories. For our study, we synthesized the axial coding results to derive core categories such as barriers to sustainability, support systems, and benefits of sustainability. With selective coding, we obtained three core categories (Table 3), resulting in a comprehensive model that encapsulates the primary factors influencing the adoption of sustainable practices in small companies, providing a clear and cohesive narrative of our findings.

Table 3 Core categories, relational structures, and their connotations derived from selective coding, own study

Core category	Typical relational structure	Connotations of the relational structure
Barriers to sustainability	Financial barriers, Knowledge gaps, Economic challenges	Financial constraints, lack of awareness and training, and economic instability hinder the adoption of sustainable practices
Support systems	Governmental support	Regulatory frameworks, incentives, and training programs from the government are crucial for promoting sustainability
Benefits of sustainability	Environmental and economic benefits	Adoption of sustainable practices leads to long-term cost savings, improved efficiency, and conservation of resources

The selective coding table provides the essence of the factors influencing the adoption of sustainable practices in small companies. It identifies three core categories: Barriers to Sustainability, Support Systems, and Benefits of Sustainability. Each category is supported by a relational structure that describes its components and their interactions. This table provides a clear and cohesive framework for understanding the complex dynamics of sustainability adoption in small companies, highlighting the challenges and the critical role of government support in facilitating sustainable practices.

Figure 3 presents a model for how small construction businesses in Angola can adopt green supply chain management (GSCM), based on our detailed analysis. This model outlines the main factors influencing the adoption of sustainable practices and is divided into three categories: barriers to sustainability, support systems, and benefits of sustainability. Barriers include high costs, limited financial resources, lack of training and awareness, and economic challenges such as instability and inflation. The model emphasizes the essential role of government support, which includes clear regulatory frameworks, training incentives, and practical guidelines to help businesses overcome these barriers. Additionally, the model highlights the benefits of adopting sustainable practices, such as long-term cost savings, improved efficiency, and conservation of natural resources, providing compelling reasons for small construction enterprises to embrace green practices.

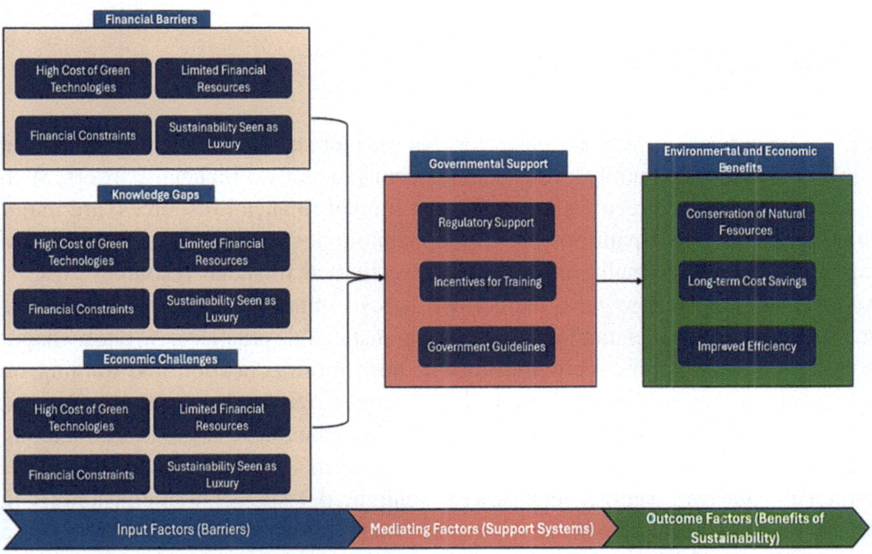

Fig. 3 Green supply chain management adoption model for small construction enterprises, own study

4.4 Theoretical Saturation Test

In our study, we conducted 12 in-depth interviews over a span of one year, from May 2023 to April 2024. The purpose of these interviews was to explore the factors influencing the adoption of green supply chain management practices among small construction companies. To ensure the robustness of our findings, we employed the concept of data saturation. For that, we began by conducting a series of initial interviews, which provided a broad understanding of the primary themes and concepts related to sustainable practices in small companies.

After each interview, we performed preliminary coding and analysis to identify emerging patterns and themes. As we conducted more interviews, we closely monitored the point at which new interviews ceased to reveal novel insights. Specifically, we looked for repetition of themes, concepts, and statements across the interviews. According to Cao (2019), when the data and information extracted by the sample are saturated, and enough theory is obtained, the test is passed. By the 10th interview, we observed that no new themes or significant information were emerging. The subsequent two interviews (11th and 12th) confirmed that the data had become repetitive, indicating that we had reached data saturation. The point of saturation was evident when additional interviews did not contribute new information, reinforcing the adequacy of our sample size.

4.5 Barriers to Sustainability

Our research found significant obstacles that prevent small construction firms from implementing green supply chain management practices. Financial barriers, such as the high costs of green technologies and limited financial resources, are major deterrents. One participant noted, "Green technologies and materials can be quite expensive, and as a small company, we have limited financial resources" (SC1). Additionally, knowledge gaps, including a lack of training and limited awareness, prevent firms from understanding and adopting sustainable practices. As one manager expressed, "Many of us, including myself, need more education and training on how to effectively implement sustainable practices" (SC1). Economic challenges, such as economic instability and inflation, further exacerbate these issues, making it difficult for small companies to prioritize long-term sustainable investments over immediate survival. Another interviewee highlighted, "The economic instability in Angola makes it difficult to invest in new, potentially costly initiatives" (SC1). These barriers align with findings from other studies in the region, highlighting the fact that many companies in the region face similar obstacles when it comes to adopting green supply chain management practices (Abdelazeem 2013; Elbarky and Elzarka 2015; Ojo et al. 2014; Tembo-Mwanaumo et al. 2023; Niemann et al. 2016).

4.6 *Support Systems*

Governmental support emerged as a critical enabler for the adoption of GSCM practices. Regulatory frameworks, incentives for training, and clear government guidelines are essential to create a supportive environment for sustainability. One participant emphasized, "A supportive policy environment from the government would provide a big boost for GSCM adoption" (SC2). The need for stronger regulations and policies that mandate or encourage sustainable practices in the construction industry was highlighted as a key factor in driving change. Incentives for training programs focused on GSCM are also crucial to build the necessary skills and knowledge within small firms. Another interviewee mentioned, "Incentives for training programs focused on GSCM would help build the necessary skills and knowledge" (SC3).

4.7 *Benefits of Green Supply Chain Management*

Despite the challenges, the adoption of GSCM practices offers significant benefits for small construction enterprises. These include long-term cost savings through improved efficiency and reduced waste, conservation of natural resources, and healthier and safer working environments. One manager pointed out, "Using energy-efficient machinery has reduced our energy consumption and operating costs" (SC6). The potential for long-term financial and operational benefits provides a strong incentive for firms to overcome initial barriers and invest in sustainability. Similarly, studies by Agyepong and Nhamo (2017), Mafini and Muposhi (2017), and Muvaka and Henry (2018) found the same benefits in their research. Furthermore, our study underscores the importance of a multifaceted approach to promoting the adoption of green supply chain management. Addressing financial constraints, enhancing knowledge and awareness, and providing robust governmental support are all critical components of this strategy. By focusing on these areas, policymakers and industry stakeholders can facilitate the transition towards more sustainable practices in the construction sector.

4.8 *Model Interpretation*

The theoretical model developed from our study encapsulates the primary factors influencing the adoption of sustainable practices in small construction enterprises in Angola. This model, depicted in Fig. 3, illustrates the interplay between barriers to sustainability, support systems, and the benefits of adopting GSCM practices.

Additionally, when compared to other local and global studies, we identified similarities with existing research on green supply chain management (GSCM) in the construction industry.

Barriers to Sustainability. Financial Barriers: High initial costs and limited financial resources restrict the ability of small firms to invest in green technologies and practices.

Knowledge Gaps: Lack of training and awareness about GSCM prevents effective implementation and adoption.

Economic Challenges: Economic instability and inflation create an environment where long-term investments in sustainability are deprioritized.

Similar to other regions where high costs of green technologies and limited financial resources complicate the adoption of green practices, the financial constraints faced by many companies in Angola are a significant barrier to the implementation of green supply chain management. For instance, studies by Elbarkouky (2013) and Ojo et al. (2014) highlight similar financial constraints in their assessments of green supply chain management in construction projects across various developing countries. Elbarkouky (2013) found that the high initial costs of green technologies deterred companies from adopting sustainable practices, despite the long-term benefits (Elbarkouky 2013). Similarly, Ojo et al. (2014) noted that limited access to financial resources and the absence of supportive financial frameworks have a negative impact on the implementation of green supply chain practices in Nigeria and South Africa (Ojo et al. 2014). These findings show the critical need for financial support and incentives to overcome the barriers to adopting green supply chain management in Angola and other developing regions.

Knowledge gaps, particularly the lack of training and awareness about green practices, can lead to resistance to change among organizations, resulting in hesitancy to adopt green practices due to a lack of understanding of their importance and benefits. In Angola, the awareness and understanding of green supply chain management (GSCM) among small construction enterprises are significantly lower than in more developed regions, highlighting the need for more comprehensive training and education in this area. However, this challenge is not unique to Angola. Studies by Ojo et al. (2014), Balasubramanian (2012), Balasubramanian and Shukla (2017), and Sourani and Sohail (2011) have identified similar issues in various developing countries. Ojo et al. (2014) highlighted that limited knowledge and awareness are major barriers to GSCM adoption in many developing nations (Ojo et al. 2014). Balasubramanian (2012) found that inadequate training and awareness programs lead to significant resistance to adopting green practices (Balasubramanian 2012). Further, Balasubramanian and Shukla (2017) emphasized the need for targeted education and training initiatives to bridge these knowledge gaps and facilitate the transition to greener practices (Balasubramanian and Shukla 2017). Sourani and Sohail (2011) also highlighted the role of government and industry bodies in enhancing awareness and understanding of GSCM through structured training programs (Sourani and Sohail 2011).

The current economic conditions in Angola, deeply affected by high inflation and post-war reconstruction challenges, add to the unique challenges faced by small construction enterprises, highlighting notable differences specific to the country.

Support Systems. Governmental Support: Strong regulatory frameworks, incentives for training, and clear guidelines from the government are essential to overcome barriers and promote GSCM adoption. One participant emphasized, "Clear regulations and guidelines from the government would help create a more supportive environment for adopting these practices" (SC4).

The support from the government is crucial for accelerating the adoption of green supply chain management practices. By establishing regulatory frameworks, offering incentives, and providing clear guidelines, governments can create a supportive environment that promotes sustainable practices. For example, the Angolan government could implement waste recycling policies and supportive measures similar to those studied by Jafari and Safarzadeh (2024), who used game theory to demonstrate how such policies promote waste recycling and resource conservation in manufacturing processes. Additionally, Wiredu et al. (2024) highlighted the positive impact of top management support, driven by government regulations, on corporate environmental performance. Their findings suggest that financial incentives and regulatory compliance can significantly enhance green practices within supply chains (Wiredu et al. 2024). In addition to that, Hou et al. (2023) further demonstrated that government subsidies could motivate businesses to adopt greener supply chains, although they noted that excessive subsidies might be counterproductive. The European Union's Green Deal and the U.S. Production Tax Credit are examples of successful government initiatives that have both catalyzed the integration of renewable energy and sustainable practices in supply chains, highlighting the importance of government support.

Benefits of Sustainability. Environmental and Economic Benefits: Adoption of sustainable practices leads to long-term cost savings, improved efficiency, conservation of natural resources, and better working conditions.

The benefits of adopting sustainable practices, such as long-term cost savings and operational efficiencies, are consistent with other studies. For instance, Agyepong and Nhamo (2017) found that green supply chain management leads to significant financial benefits and improved efficiency, reinforcing our findings on the positive impacts of sustainable practices.

Our model adds significant value by specifically addressing the socio-economic and environmental challenges of Angola's small construction enterprises. It integrates targeted financial incentives, capacity-building initiatives, and robust regulatory frameworks. This approach not only helps overcome unique barriers faced by small Angolan firms in post-war economic instability scenarios and in an economy that is highly affected by high inflation but also emphasizes the critical role of governmental support in facilitating the adoption of green practices in the country. By promoting long-term cost savings, operational efficiency, and resource conservation, our model provides a practical and replicable roadmap for other developing nations with similar conditions, enhancing sustainability on a broader scale.

4.9 *Level of Awareness and Understanding of GSCM Concepts*

The level of awareness and understanding of green supply chain management practices within small construction firms in Angola was found to be relatively low. Many participants expressed limited knowledge about GSCM principles and practices. This lack of understanding is a significant barrier to the adoption of sustainable practices. For instance, several interviewees admitted that they had not received any formal training on GSCM and were not fully aware of its potential benefits. This highlights the need for targeted educational and training programs to build capacity and foster a deeper understanding of GSCM among small construction enterprises.

Our theoretical model is important and practical. It addresses a gap in green supply chain management literature by focusing on small construction businesses in Angola, a rarely studied area, using a grounded theory approach reflecting real-world experiences. This model highlights key barriers and support systems, emphasizing the need for government support, financial incentives, and training programs. By showcasing benefits like cost savings, improved efficiency, and environmental conservation, the model encourages small construction firms to adopt sustainable practices, enhancing their competitiveness and contributing to broader environmental and social goals.

5 Conclusion

This study aimed to explore the adoption of green supply chain management (GSCM) practices among small construction enterprises in Angola, focusing on the underlying mechanisms and contextual factors shaping this adoption. Using a grounded theory approach, we conducted in-depth interviews with professionals from 12 small construction firms to understand their experiences, perceptions, and operational strategies related to GSCM integration. Green supply chain management is a sustainable approach that aims to safeguard the environment.

Our findings revealed significant barriers to adopting GSCM practices, including financial constraints, knowledge gaps, and economic instability. High costs associated with green technologies and limited financial resources were major obstacles. Additionally, a lack of training and awareness about GSCM principles hindered effective implementation. Economic challenges, such as inflation and market volatility, exacerbated these issues, making it difficult for small enterprises to prioritize long-term sustainability over immediate survival. However, we identified several critical support systems that can facilitate GSCM adoption. Governmental support, through regulatory frameworks, financial incentives, and training programs, was essential for creating a conducive environment for sustainability. Participants emphasized the need for stronger regulations and policies that encourage or mandate sustainable practices in the construction industry.

The benefits of adopting GSCM practices included long-term cost savings, improved efficiency, and conservation of natural resources. These benefits provide compelling reasons for small construction enterprises to overcome initial barriers and invest in sustainability. Moreover, by leveraging support systems and addressing financial and educational barriers, it is possible to drive a transformative shift towards environmentally responsible construction practices.

The theoretical model we developed encapsulates the primary factors influencing the adoption of sustainable practices in small construction enterprises in Angola. This model can inform strategies in other emerging economies facing similar challenges, highlighting the importance of robust policy support and targeted interventions to promote sustainable practices.

Our study contributes significantly to the body of knowledge in several ways. Firstly, by exploring GSCM adoption in small construction businesses in Angola, it addresses a critical research gap in an underexplored context, providing valuable insights into the unique challenges and dynamics faced by these firms. Secondly, the development of a grounded theory model offers a theoretical foundation for understanding the factors influencing GSCM implementation in similar contexts, informing future research and practical interventions. Regionally, the findings can be adapted and applied to other African countries facing similar challenges, offering a framework for promoting sustainable practices across the continent. The insights gained from this study are not only relevant to Angola but also contribute to the broader discourse on sustainability in developing economies, enriching the global understanding of GSCM adoption. However, this study has some limitations. The relatively small sample size of 12 companies may not fully capture the diversity of experiences within the broader industry. Additionally, the geographical concentration in Luanda, Benguela, and Huambo may limit the applicability of the findings to other regions of the country. Also, the need to anonymize participants to protect confidentiality could restrict the contextualization of specific insights.

Finally, to facilitate the adoption of sustainable practices in the construction industry, we recommend targeted financial incentives, educational programs, and stronger governmental policies. Additionally, collaboration between government agencies, industry associations, and educational institutions is also crucial for creating a supportive ecosystem for the adoption of green supply chain management.

Appendix A

See Table 4.

Table 4 Open code result, own elaboration

Initial category	Initial concept	Primitive statement
Cost challenges	High cost of green technologies	SC1. Green technologies and materials can be quite expensive, and as a small company, we have limited financial resources
		SC1. High initial costs associated with sustainable technologies and materials can be an obstacle at times like this
	Financial constraints	SC2. The upfront costs of green technologies can be prohibitive. We operate on tight margins, so sustainable options often seem out of reach
		SC8. Investing in sustainable practices can seem like a luxury we cannot afford, given our tight budgets
	Sustainability seen as luxury	SC10. Many see sustainability as a luxury rather than a necessity, especially when immediate survival is at stake
		SC4. Sustainable practices are often perceived as costly and complex, which can deter small companies with limited budgets
Lack of knowledge	Lack of training	SC1. Many of us, including myself, need more education and training on how to effectively implement sustainable practices
		SC7. Education and training are crucial for promoting GSCM. They help build awareness and understanding of the benefits and practices associated with GSCM
	Limited awareness	SC4. Many small business owners, like myself, are focused on immediate operational concerns and may not have the time or resources to learn about new practices
		SC5. There is a general lack of awareness and education about GSCM in the construction sector
	Focus on immediate concerns	SC12. Sustainability hasn't really been a top priority when we're struggling with issues like accessing affordable materials, reliable power, transportation challenges, and stretched budgets
		SC2. Most construction companies here are just focused on completing projects efficiently and profitably using conventional methods
Economic instability	Limited financial resources	SC7. Limited financial resources make it hard to prioritize long-term sustainable practices over immediate financial survival
		SC6. High initial costs associated with sustainable technologies and materials can be prohibitive

(continued)

Table 4 (continued)

Initial category	Initial concept	Primitive statement
	Inflation	SC12. Our industry has been severely impacted by ups and downs in oil prices, exchange rate fluctuations, and inflation
		SC5. Economic instability and inflation make it difficult to prioritize long-term sustainable practices
	Economic fluctuations	SC1. The economic instability in Angola makes it difficult to invest in new, potentially costly initiatives
		SC9: "Economic fluctuations make it challenging for us to invest in sustainable practices
Government support	Regulatory support	SC2. A supportive policy environment from the government would provide a big boost for GSCM adoption
		SC9. Stronger regulations and policies that encourage or mandate sustainable practices in the construction industry could help drive change
	Incentives for training	SC3. Incentives for training programs focused on GSCM would help build the necessary skills and knowledge
		SC12. Training on modern sustainable methods, materials, and equipment would help build skills and buy-in
	Government guidelines	SC4. Clear regulations and guidelines from the government would help create a more supportive environment for adopting these practices
		SC10. A technical assistance program from the government could help small construction companies transition to greener operations
Environment and economy	Conservation of natural resources	SC6. Using energy-efficient machinery has reduced our energy consumption and operating costs
		SC8. Sustainable practices can lead to healthier and safer working environments for employees and better living conditions for the community
	Long-term Cost Savings	SC2. Although there might be initial costs, sustainable practices can lead to cost savings in the long run through improved efficiency and reduced waste
		SC3. Reducing waste and increasing efficient use of resources directly translates into cost savings over the long run

(continued)

Table 4 (continued)

Initial category	Initial concept	Primitive statement
	Improved efficiency	SC6. Using energy-efficient machinery has reduced our energy consumption and operating costs
		SC11. Adopting energy-saving equipment and sustainable materials could translate into meaningful cost savings and productivity gains

References

African Development Bank (2018) Angola Economic Outlook. https://www.afdb.org/fileadmin/uploads/afdb/Documents/Publications/African_Economic_Outlook_2018_-_EN.pdf. Access 24 Jan 2024

African Development Bank (2020) African Economic Outlook 2020. African Development Bank, Abidjan. https://www.afdb.org/sites/default/files/documents/publications/african_economic_outlook_2020-en.pdf

Agyepong AO, Nhamo G (2017) Green procurement in South Africa: perspectives on legislative provisions in metropolitan municipalities. Environ Dev Sustain 19(6):2457–2474

Ali M, Din MGU, Cheema SM (2023) Integrating sustainability into construction engineering projects: perspective of sustainable project planning. J Dev Soc Sci 4(I)

Aragón-Correa JA, Hurtado-Torres N, Sharma S, García-Morales VJ (2008) Environmental strategy and performance in small firms: a resource-based perspective. J Environ Manage 86(1):88–103

Atkinson R, Flint J (2001) Accessing hidden and hard-to-reach populations: snowball research strategies. Soc Res Updat 33(1):1–4

Azevedo SG, Carvalho H, Cruz Machado V (2011) The influence of green practices on supply chain performance: a case study approach. Transp Res Part E: Logist Transp Rev 47(6):850–871

Balasubramanian S (2012) A hierarchical framework of barriers to green supply chain management in the construction sector. J Sustain Dev 5(10):15–27

Balasubramanian S, Shukla V (2017) Green supply chain management: an empirical investigation on the construction sector. Supply Chain Manag: Int J 22(1):58–81. https://doi.org/10.1108/SCM-07-2016-0227

Barbosa F, Woetzel J, Mischke J (2017) Reinventing construction: a route of higher productivity. McKinsey Global Institute

Begu L, Vasilescu M, Stanila L, Clodnitchi R (2018) China-Angola investment model. Sustainability 10(8):2936

Benmamoun Z, Hachimi H, Amine A (2017) Green logistics practices. In: 2017 International Renewable and Sustainable Energy Conference (IRSEC), pp 1–5

Bitsch V (2005) Qualitative research: a grounded theory example and evaluation criteria

Bühler MM, Hollenbach P, Michalski A, Meyer S, Birle E, Off R, Lang C, Schmidt W, Cudmani R, Fritz O, Baltes G, Kortmann G (2023) The industrialisation of sustainable construction: a transdisciplinary approach to the large-scale introduction of compacted mineral mixtures (CMMs) into building construction. Sustainability 15(13):10677

Cao Q, Sarker MNI, JiangYan SJS (2019) Model of the influencing factors of the withdrawal from rural homesteads in China: application of grounded theory method. Land Use Policy 85:285–289

Corbin J, Strauss A (2008) Qualitative research. Techniques and procedures for developing grounded

Del Passo OF (2019) INDÚSTRIA DA CONSTRUÇÃO CIVIL E BLOCO NO PODER (2011–2016). REVES - Revista Relações Sociais 2(3):0433–0446

Dhawan R, Malik HK, Dorranian D (2023) Applications of sustainable materials. In: Ajay, Parveen, Ahmad S, Sharma J, Gambhir V (eds) Handbook of sustainable materials: modelling, characterization, and optimization, 1st edn. CRC Press, pp 301–322

Dunne C (2011) The place of the literature review in grounded theory research. Int J Soc Res Methodol 14(2):111–124

Elbarkouky MMG, Abdelazeem G (2013) A green supply chain assessment for construction projects in developing countries. In: WIT Transactions on Ecology and the Environment, pp 1331–1341

Elbarky S, Elzarka S (2015) A green supply chain management migration model based on challenges faced in Egypt. In: 2015 International Conference on Industrial Engineering and Operations Management (IEOM), pp 1–8

Tembo-Mwanaumo E, Mutono-Mwanza B, Siaminwe L (2023) Green supply chain management practices and effects on organizational performance in the Zambian beverage manufacturing industry. Logist Oper Manag

Fahimnia B, Sarkis J, Davarzani H (2015) Green supply chain management: a review and bibliometric analysis. Int J Prod Econ 162:101–114

Glaser BG, Strauss AL (1967) The discovery of grounded theory: strategies for qualitative research. Aldine Publishers, Chicago

Glaser BG, Strauss AL (2017) The discovery of grounded theory: strategies for qualitative research, 1st edn. Routledge

Government of Angola (2018): National Development Plan 2018–2022. Retrieved from https://www.governo.gov.ao/angola-2025/pnd-2018-2022/

Hervani AA, Helms MM, Sarkis J (2005) Performance measurement for green supply chain management. Benchmarking: Int J 12(4)

Hou L, Zhang Y, Wu C, Song J (2023) Improving the greenness of enterprise supply chains by designing government subsidy mechanisms: based on prospect theory and evolutionary games. Front Psychol 14

International Energy Agency (2019): Tracking Buildings. Retrieved from https://www.iea.org/reports/tracking-buildings-2019

James WM (2020) A political history of the civil war in angola 1974–1990. In: James Iii WM 1st edn, Routledge

Jafari H, Safarzadeh S (2024) Effects of governmental supportive policies on waste management for two substitutable products made of virgin and waste materials: A game-theoretic approach. Waste Manag Res

Jeong, Y. (Ed.).: Sustainable Construction Materials: From Paste to Concrete. MDPI (2023).

Mafini C, Muposhi A (2017) The impact of green supply chain management in small to medium enterprises: cross-sectional evidence. J Transp Supply Chain Manag 11

Mingqiang Z, Zuxu Z (2011) Green supply chain management in construction industry. In: Dai M (Eds), Innovative Computing and Information, Vol 232, Springer Berlin Heidelberg, pp 81–86

Mutingi M, Mapfaira H, Monageng R (2014) Developing performance management systems for the green supply chain. J Remanufacturing 4(1):6

Muvaka MK, Henry B (2018) Green supply chain practices as a supply chain performance tool in state corporations in Kenya, case study of Kenya electricity generating company limited. Int J Soc Sci Manag Entrep (IJSSME) 1(1)

Niemann W, Kotze TG, Adamo F (2016) Drivers and barriers of green supply chain management implementation in the Mozambican manufacturing industry. J Transp Supply Chain Manag 13(1):977–1013

Noy C (2008) Sampling knowledge: the hermeneutics of snowball sampling in qualitative research. Int J Soc Res Methodol 11(4):327–344

Ojo E, Mbowa C, Akinlabi ET (2014) Barriers in implementing green supply chain management in construction industry. In: International Conference on Industrial Engineering and Operations Management, pp 7–9

Ojo E, Mbohwa C, Akinlabi E (2014) Green supply chain management in construction industries in South Africa and Nigeria. Int J Chem Environ Biol Sci (IJCEBS) 2(2):146–150

Pautasso D (2010) A Economia Política Internacional da China para Angola e os Caminhos da Transição Sistêmica. SÉCULO XXI: Revista de Relações Internacionais - ESPM-POA 1(1)

Ra S, Chin B, Liu A (2015) Challenges and opportunities for skills development in Asia: changing supply, demand, and mismatches. Asian Development Bank
Rao P, Holt D (2005) Do green supply chains lead to competitiveness and economic performance? Int J Oper Prod Manag 25(9):898–916
Sarkis J (2003) A strategic decision framework for green supply chain management. J Clean Prod 11(4):397–409
Sarkis J, Zhu Q, Lai K (2011) An organizational theoretic review of green supply chain management literature. Int J Prod Econ 130(1):1–15
Simpson JA, Collins WA, Tran S, Haydon KC (2007) Attachment and the experience and expression of emotions in romantic relationships: a developmental perspective. J Pers Soc Psychol 92(2):355–367
Sosnowski P (2002) Green concepts in the supply chain. Logforum 18 (1):2. https://doi.org/10.17270/J.LOG.2022.680
Sourani A, Sohail M (2011) Barriers to addressing sustainable construction in public procurement strategies. In: Proceedings of the Institution of Civil Engineers, Engineering Sustainability 164(4):229–237
Søreide T (2011) Projectos de Construção Pública—Angola. Necessidade de fortificar as barreiras contra a corrupção. Angola Brief 1(6)
Srivastava SK (2007) Green supply-chain management: a state-of-the-art literature review. Int J Manag Rev 9(1):53–80
Turin DA (1973) The Construction Industry: Its Economic Significance and its Role in Development. UCERG, London
UNEP-SBCI (2023): http://www.unep.org/sbci/AboutSBCI/Background.asp. Accessed 28 Aug 2023
UNEP (2013): Green Economy and Trade—Trends, Challenges and Opportunities. United Nations Environment Programmed
United Nations Environment Programme (2016) Global Status Report 2016: Towards Zero-Emission Efficient and Resilient Buildings. UNEP, Nairobi. Available at https://wedocs.unep.org/handle/20.500.11822/10618
United Nations Environment Programme (UNEP) (2021) Sustainable Public Procurement Implementation Guidelines, 2nd edn. UNEP, Nairobi. Available at https://www.unep.org/resources/publication/second-edition-uneps-sustainable-public-procurement-guidelines
United Nations Environment Programme (2023): Angola: Country Environmental Analysis. https://www.un.org/sustainabledevelopment/blog/2021/02/angola-country-environmental-analysis/. Access 7 Nov 2023
Walker D, Myrick F (2006) Grounded theory: an exploration of process and procedure. Qual Health Res 16(4):547–559
Walker H, Di Sisto L, McBain D (2008) Drivers and barriers to environmental supply chain management practices: lessons from the public and private sectors. J Purch Supply Manag 14(1):69–85
Wanda F, Oya C, Monreal B (2023) Building angola: a political economy of infrastructure contractors in post-war Angola. J South Afr Stud 49(1):25–47
Wiredu J, Yang Q, Sampene AK, Gyamfi BA, Asongu SA (2024) The effect of green supply chain management practices on corporate environmental performance: does supply chain competitive advantage matter? Bus Strateg Environ 33(3):2578–2599
Worldometer (2022): CO2 Emissions by Country—Worldometer. CO2 Emissions by Country. https://www.worldometers.info/co2-emissions/co2-emissions-by-country/. Access 25 Nov 2023
Ye M, Ye K (2013) Game relationships between the stakeholders of Angola mode-based construction projects. ICCREM 2013:438–447
Yin RK (2009) Case study research: design and methods, Vol 5, Sage
Zhu Q, Sarkis J, Lai K (2007) Initiatives and outcomes of green supply chain management implementation by Chinese manufacturers. J Environ Manage 85(1):179–189

Use of IT Tools to Support Digitalisation and Optimalisation in Companies and Supply Chains

Iterative Simulations Supported by GIS-Class Tools Based on a Supply Chain for Liquid Products

Damian Dubisz, Chawis Boonmee, Rafał Rokicki, and Paulina Golińska-Dawson

Abstract Supply chains have evolved into complex, multi-layered models that involve a variety of global stakeholders and participants and operate from geographically dispersed locations. Therefore, it is important to apply the most sufficient management methods and tools to support both the effective selection of appropriate solutions and the configuration of global supply chains (GSCs). The literature review revealed the potential of combining traditional GSCs management methods with an iterative approach. The empirical part, related to the analysis of efficiency ratios between iteration results, was carried out using actual data from a liquids-based supply chain. The research focused on identifying changes in the transport efficiency of liquid material flows. Five to-be iterations and one as-is iteration were proposed for model calibration. It was validated that a change in planning method of delivery points along routes and consolidation of deliveries can affect supply chain efficiency. Changing the functionality of process participants and the regionalisation of delivery areas was also found to improve transport efficiency by up to 22.37%, depending on the other parameters changed in each of the different iterations. The study was based on anonymised historical data from a liquid-oriented transport company. The analysis was carried out using ArcGIS tool with the Network Analyst add-on. The result of the research provides valuable conclusions that lead to the recommendations that may be used when planning a GSCs design. Based on the detailed calculations, the most sufficient control parameters can be pointed out for an implementation within existing supply chains in order to obtain a more sustainable logistic model.

D. Dubisz (✉) · P. Golińska-Dawson
Faculty of Engineering Management, Poznan University of Technology, Poznań, Poland
e-mail: damian.dubisz@pit.lukasiewicz.gov.pl

P. Golińska-Dawson
e-mail: paulina.golinska@put.poznan.pl

C. Boonmee
Department of Industrial Engineering, Faculty of Engineering, Chiang Mai University, Chiang Mai, Thailand
e-mail: chawis.boonmee@cmu.ac.th

D. Dubisz · R. Rokicki
Łukasiewicz Research Network – Poznań Institute of Technology, Poznań, Poland
e-mail: rafal.rokicki@pit.lukasiewicz.gov.pl

A. Kolinski and M. Adamczak (eds.), *Digitalisation of the Greening Supply Chain*, EcoProduction, https://doi.org/10.1007/978-3-031-88918-9_5

Keywords Iterative approach · Supply chain management · Supply chain parameters · Sustainable supply chains · Logistic model improvement · Logistic simulations

1 Introduction

Supply chains become highly complex structures that require proper modelling to maintain required level of efficiency (Mogale et al. 2022; Arena et al. 2021; Cai et al. 2009; Kara et al. 2007). The multilayer structure of modern sustainable supply chains is a significant factor influencing their effectiveness (López and Ruiz-Benítez 2020). It is essential to consider the numerous mutual interconnections between process participants in terms of their functionality and operational range. Technological developments allow efficiency analyses of a variety of supply chain organisation options to be carried out under simulation conditions. This research aims to verify the potential to change supply chain efficiency by modifying the control parameters of the model. Within the research, the efficiency verification has been carried out for several alternative iterations. The main concern while determining the direction of management changes within the supply chain is the need to refer to historical data and efficiently verify their validity. For this purpose, it is necessary to use real company's historic data and the conclusion must refer to the analysis of the performance indexes evaluated in each iteration. The research carried out has the character of a case study, based on real historical data of a company handling the transport of liquid products.

2 Literature Review

The literature focused on process management improvement within global supply chains points to the important role of a number of factors in determining both the efficiency and performance of those. Research by Dubisz et al. (2022a) points to the significance of fleet parameters, the energy intensity of the transport process, and the route parameters. The research of the expert panel points to the importance of these three areas among transport process. It has been identified that proper management of a heterogeneous transport fleet can be improved by taking into consideration a number of parameters related to the gross vehicle mass (GVM), as well as age of vehicles (Dubisz et al. 2022b). The transport efficiency parameters are primarily related to both the functionality trust among the of supply chain participants (Ghosh and Fedorowicz 2008). By properly modelling the links between the key elements of the supply chain, the efficiency of the entire transport chain can be influenced. Furthermore, a corporate Environmental, Social, and Governance (ESG) approach, which takes into consideration all key elements that often impact the local community wellbeing and processes carbon footprint levels during the GSCs modelling. Such an approach can help align the supply chain with all the criteria characteristic of sustainable models (Gao et al. 2024). A good approach to supply chain modelling is to exploit available tools to specify not only the functionality of participants, but also their geographical distribution.

This can be supported by a dedicated algorithm considering all important elements of Supply Chain (SC) infrastructure (Tao et al. 2018). However, traditional methods of determining the centre of gravity of locations for specific warehouses, cross-docks, loading hubs and other SC elements can also be adopted (Martyushev et al. 2023). The nature of complex global supply chains based on dispersed suppliers and customers around the world is indicated by Kannegiesser and Günther (2014). Global supply chains must take into account the specific characteristics of the region in which they are located and where some of the core processes are undertaken. This is directly related to corporate social responsibility at the local level (Chen et al. 2024). Significant distances between participants in the process indicate an important aspect of the distance parameter. Thus, the significance of the distance parameter proposed by Dubisz et al. (2022a) as well as fleet and energy intensity become even more important. Regardless of the type of products that are being handled, each of the management and decision-making approaches within the supply chain should be based on a set of efficiency indicators dedicated to the specifics of the supply chain (Cai et al. 2009). Therefore, it seems reasonable to propose in each case its own performance indicators related to both the physical form of the product and, the distance or energy required to perform the process. The importance of control parameters that influence the efficiency of the entire supply chain has also been identified in studies of the electrical and electronic equipment (EEE) industry (Werner-Lewandowska et al. 2024). The change from a traditional supply chain model to a sustainable model based on circularity concepts points to up to 30 essential parameters that define efficiency. Key groups of factors relate to recovery, recycling, remanufacturing and repair scenarios. The research also shows that a well-functioning service model for the flow of goods is based on good planning of logistics processes and transformational leadership in the courier services (Sayan and Sürücü 2024).

The aim of current logistics concepts is to obtain the best ratio between resource usage and results of related logistic process outcome (Cai et al. 2009). This approach has become even more important in the context of the current crises affecting global supply chains (Durugbo and Al-Balushi 2023). It would be beneficial to introduce lean management elements, which become beneficial especially in production-oriented companies. Another good example might be an agile approach that supports the introduction of tailored to the actual needs. Limited access to the resources required for essential transport processes requires increasing the sustainability of supply chains. An approach based on agile concepts makes it possible to achieve improved efficiency in some cases by introducing greater flexibility within the supply chain, in particular when customer service levels and environmental aspects are measured (Stachowiak et al. 2013). Despite of the method chosen, efficiency increase can be achieved by introducing various efficiency indicators that support constant evaluation of the supply chain model. One of them would be a carbon footprint indicator on which a wide range of logistics management concepts deeply rely (Cadarso et al. 2016).

The integration of artificial intelligence (AI) into geographic information systems (GIS) holds considerable potential for enhancing the evaluation of supply chain efficiency through the utilisation of the iteration approach (Emami and Marzban 2023). This integration becomes particularly advantageous when dealing with large datasets. Moreover, the integration of Machine Learning (ML) algorithms and regression techniques is likely to enhance the precision of the results, aligning more closely with the requirements of the supply chain management process (Almalki et al. 2021). The creation of predictive models has been shown to influence the resistance level of the entire logistics model, which is also crucial requirement for stakeholders.

A practical approach to management level decision making can be the use of a multi-criteria decision matrix. The ability to consider a range of relevant environmental, cost, and fleet availability factors provides the opportunity to base decision processes on specific parameters to which a precise level of importance is assigned (Zając et al. 2024). The Multi-Criteria Selection Matrix can be used successfully in an iterative simulation approach considering single and multiple sequences (Gelman and Rubin 1992). When comparing the results of selecting a freight forwarder using the Multi-Criteria Decision-Making (MDCM) tool, it is possible to identify the most advantageous supply chain configuration and select the most efficient supply chain participants to carry out the process (López and Ruiz-Benítez 2020; Zając et al. 2024). An important element of current supply chains is not only social responsibility, as part of the ESG approach, but also competence management of members of the logistics team. Corporate Social Responsibility for Competency Management (CSRFCM) is an important element to consider while managing the supply chain (Goliński and Szafrański 2019). However, in an iterative approach, it is not possible to verify changes in this parameter. Further research could be conducted on the index of CSRFCM and attempts to parameterise it. Then, such an evaluation using the MCDM tool would be possible.

The assumptions of the circular economy include minimising the environmental impact of logistics processes while maintaining economic growth (Golinska-Dawson 2020). The zero-waste approach that exists in production and processing processes can also be applied to the modelling of transport processes. Minimising environmental impact and increasing efficiency can be achieved through a bottom-up approach while introducing sustainable practices within the SC. Incorporation of innovative technological solutions has been shown to lead to a transition to a more sustainable logistics model. An environmental approach suitable for application in an iterative approach refers to the verification of CO_2 emissions of transport processes. However, research carried out among stakeholders in the chemical industry by Cichosz and Pluta-Zaremba (2019) indicates obstacles to the application of this type of management due to the lack of sufficient legal regulations and free tools to support carbon footprint evaluations of transport processes in companies.

An interesting perspective on increasing the level of supply chain resilience through the use of artificial intelligence in modelling of SC's design, process participant selection and its configuration is provided by Singh et al. (2024) in their research. It was found that the impact of disruptions depends on the level of transparency of the organisation, the availability of data, and the degree of digitisation of processes, based on a study of the experts panel in the supply chain and the correlations between the elements of the SCs.

The research on the iterative analysis of alternative transport logistics organisation options shows that a number of factors should be taken into account. At the same time, the literature review revealed a lack of research on the importance of individual parameters. In order to fill this gap, this paper examines the importance of individual groups of parameters.

3 Research Methodology

Adopting an appropriate framework for this research was the first step in analysing the potential to increase distribution efficiency in alternative supply chain organisation scenarios. The effectiveness of alternative supply chain routes can be examined by comparison of specific efficiency indicators (Gligor and Holcomb 2012). However, each approach is related to different aspects of the assessment. Lack of sufficient research on broader models that take into account the multi-faceted management of GSCs (Cambero and Sowlati 2014). The iterative approach allows for the verification of the impact of a number of different control parameters, which are clearly defined within the analysed supply chains. Consequently, it is possible to determine the precise differences in efficiency levels between multiple iterations. Gelman and Rubin approach (1992) supporting decision-making process related to multiple actors and various supply chain control parameters and set ups allows to point out the most efficient solution adaptable within existing SC.

The objective of the research was to determine the extent of the data that would be required for the analyses. Subsequently, the collected information was subjected to analysis, and the data format was adapted to align with the specific requirements of the analytical tool employed in the research. An indicator and process analyses were conducted using the primary data. This resulted in the formulation of preliminary conclusions and findings. The findings of the analysis have been employed in the formulation of practical and managerial implications. In the final stage of the research, the limitations of the study and potential areas for further investigation were identified.

In accordance with the specific characteristics of the distribution model of liquid products and the methodology employed in the research, the following key elements were taken into account:

- Delivery date dependent on variable customer preferences,
- Quick price change dynamics among liquids suppliers,

- Limited types of liquid products available from suppliers at loading points.

In consideration of the above parameters, it was deemed essential to develop an additional computational model to facilitate the evaluation of route efficiency in relation to the purchase price of the liquid products and the distance between loading and unloading points along the route. The model was developed and integrated into the ArcGIS tool with the Network Analyst addon.

In order to identify alterations in distribution efficiency in subsequent iterations of the simulation conducted with the ArcGIS Pro Network Analyst tool (Version 3.3.1), it was necessary to map the current state of distribution. Therefore, a 0-0 iteration (as-is) was created, which corresponded to the historical data of the analysed enterprise with regard to efficiency parameters, the use of quantities and types of means of transport, issued litres of transported liquids. Further to-be simulations were carried out after the as-is iteration was successfully validated. The observed changes in the efficiency level in each iteration were related to the baseline iteration 0-0, thus enabling verification of the dynamics and direction of the efficiency changes in each of the presented iterations.

The objective of the simulations is to provide insight into the impact of modifications to the distribution model on the efficiency of the implemented distribution processes. The main goal of the simulation was to obtain answers for following questions:

RQ1: How the introduced change in the distribution model affects the efficiency of the implemented distribution processes?

RQ2: Whether there is a negative impact of the introduced change and an increase in the efficiency of the liquid products distribution process?

RQ3: Which changes should be made to the liquid distribution model to increase its efficiency and which changes should be avoided?

The results are presented in the subsequent sections.

4 Specification of the Analysed Liquid Product Supply Chain

The research was based on primary data collected from a company whose core business involves the logistics processes of liquid products. The company is structured hierarchically and is divided into two distinct branches, each of which serves a specific group of customers within the same region. The company provides transportation services for liquid products within the Polish market mainly, however supplies are imported from all over the world. This turns analysed SC into GSCs model. As a consequence of the distribution of customers, financial purposes, suppliers, and purchasing habits, the supply chain was divided into two separate companies. In the research, the two companies are referred to as Company 1 (C1) and Company 2 (C2). In the existing logistics model, the delivery of products to customers for Companies C1 and C2 is conducted as two separate operations for organisational reasons. Consequently, each tank vehicle is only loaded with goods belonging to Company C1 or C2. In the existing logistics model, a number of liquid product categories are identified for transportation, which are anonymised in this paper for confidentiality reasons. The relevant products are referred to as "liquid product no. 1–5." In view of the extensive number of routes completed during the year, the decision was made to use data from two weeks of process operation to ascertain the level of efficiency across the various iterations. The data used in this research covers the operational period of the third quarter of 2024 (Q3Y24) 0.4. Basic simulation assumptions and supply chain parameters. The identification of the primary parameters of the distribution process is a crucial step in the development of a simulation environment. In order to gain a deeper understanding of the specificities of handling liquid products, an in-depth examination of the key logistic parameters was conducted to gain insight into the unique aspects of those products transportation. The findings of this literature-review based analysis have led to the identification of the key elements of the transport processes within supply chain that has to be considered in iterative oriented simulations. All key GSCs parameters identified are listed in Table 1.

The table refers to research that identifies the elements of the supply chain that influence the efficiency of the logistics process. For selected parameters, reference is made to their embedding in the supply chain and their potential significant impact on efficiency within iterative simulations.

Following the completion of the review, the key parameters of the logistics model were incorporated into the ArcGIS simulation model. As a result of the parameter revision, all key elements of the analysed company's supply chain of the company analysed were reflected in the iterative simulations.

At the initial stage of the research, an analysis was conducted to evaluate the current state of the company's logistics model. The definition of the type of document employed in the logistics handling process was the foundation of a proper understanding of mutual interconnections within the supply chain. Subsequent to an analysis of the company's primary data, the incoming and outgoing documents were identified. The diagram presented in Fig. 1 illustrates the relationship between the various elements of the company's supply chain.

With reference to the identified literature on iterative approach and taking into account the structure and data quality of the company under analysis, a dedicated computational logic was proposed. An important factor was the possibility of comparing the results between iterations. Hence, the need to define a common parameter to compare was essential. A simplified formula presented below is applied during the further evaluation assessment. The following Truck Efficiency (TE) and Overall Iteration Efficiency (OIE) indicators have been proposed.

$$TE_{\left(\frac{litres}{kilometers}\right)} = \frac{\sum litres_{(products\,no.\,1-5)}}{\sum distance_{(Kilometes\,per\,vehicle)}}$$

Table 1 Key elements of the transport processes within supply chain to be reflected in iteration based computational model

Parameter	Characteristics of the parameter as indicated in the literature
Types of transport documentation accompanying events present within the liquid products distribution process	Importance of introduction of a proper types of documentation supporting liquid chemicals flow has been outline by Bruinen De Bruin et. al. (2007). Therefore, this parameter has been took into consideration within created computational iterative model. The types of documents used also affect the quality of information transferred between supply chain actors, especially in logistics models based on outsourcing of services based on 4PL concepts (Kolinski et al. 2020)
Product groups	It has been outlined in quantitative analysis how adequate management interfere with supply chain efficiency (Li and O'Brien 2001). Multiple objective optimisation model may be beneficial to obtain performance improvement within Green Supply Chains (GrSC)
Mutual interdependencies—The flow of the goods within the supply chain determined by SC participants	It has been identified that SC design influence the overall efficiency of supply chain despite of the product types handled (Garcia and You 2015)
Scope of operations of supply chain participants	Research conducted in Australia and Malaysia present a new management approach for supply chain management. In depth focus on scope of operation of each participants is essential and has the key role in efficiency level modelling (Khalfan et al. 2014)
Functional scope of process participants—areas and processes for which SC participants are responsible for	Identification of specific tasks and defining precise functions of participants within the existing supply chain is beneficial to obtain a better efficiency level according to due to specialisation in the execution of a specific process. (Garcia and You 2015; Abdulla and Musa 2021)
Means of transport—considering their capacity and general characteristics for carrying out loading and unloading (e.g. unloading time, possibility of bottom or top loading, equipping vehicles with a set of pumps to speed up the process)	It has been observed that properly equipped means of transport can significantly improve the efficiency of the logistics process especially in terms of liquid products distribution. An example of an additional tank truck with pumps can reduce the time required to unload the liquids at the delivery site (Rakhimov et al. 2021)

(continued)

Table 1 (continued)

Parameter	Characteristics of the parameter as indicated in the literature
The locations of the suppliers and their specific characteristics	Geographical spread of supply chain participants is a critical element in reducing transport emissions and influencing the overall effectiveness of a supply chain. Accurately defining the role of global suppliers and their location is the foundation of a well-managed supply chain (Egri et al. 2021)
The characteristics of order processing of liquid products among the company's customers	Appropriate order management could be profitable for overall process efficiency, Acero et al. (2019) point out that the flexibility of orders handling within supply chain plays a key role in improving its resilience
Material characteristics	It has been shown that the physical characteristics of the liquids transported have an impact on the processing methods and can influence the overall SC design concept (Gao et al. 2020)

Source own study

Truck Efficiency—Understood as the level of efficiency of transportation processes simulated within a specific iteration related to the specific mean of transport. The higher the value of the indicator, the greater the efficiency of each specific truck.

$$OIE_{\left(\frac{litres}{kilometers}\right)} = \frac{\sum litres_{(products\ no.\ 1-5)}}{\sum distance_{(Kilometes\ per\ fleet)}}$$

Overall Iteration Efficiency—The overall efficiency of transportation processes simulated within a specific iteration is related to the entire fleet of trucks. The higher the value of the indicator, the higher the efficiency.

The proposed formulas have been used as the logical basis for the verification of the overall efficiency of each iteration within the computational model.

4.1 Essential Elements of Primary Data

To facilitate the analytical process, it is essential to ensure that the input data accurately reflect the characteristics and requirements of the logistics operations under examination. The input data set included the following information:

- The point of dispatch, including the location, time window, type of liquid products loaded, and loading time.

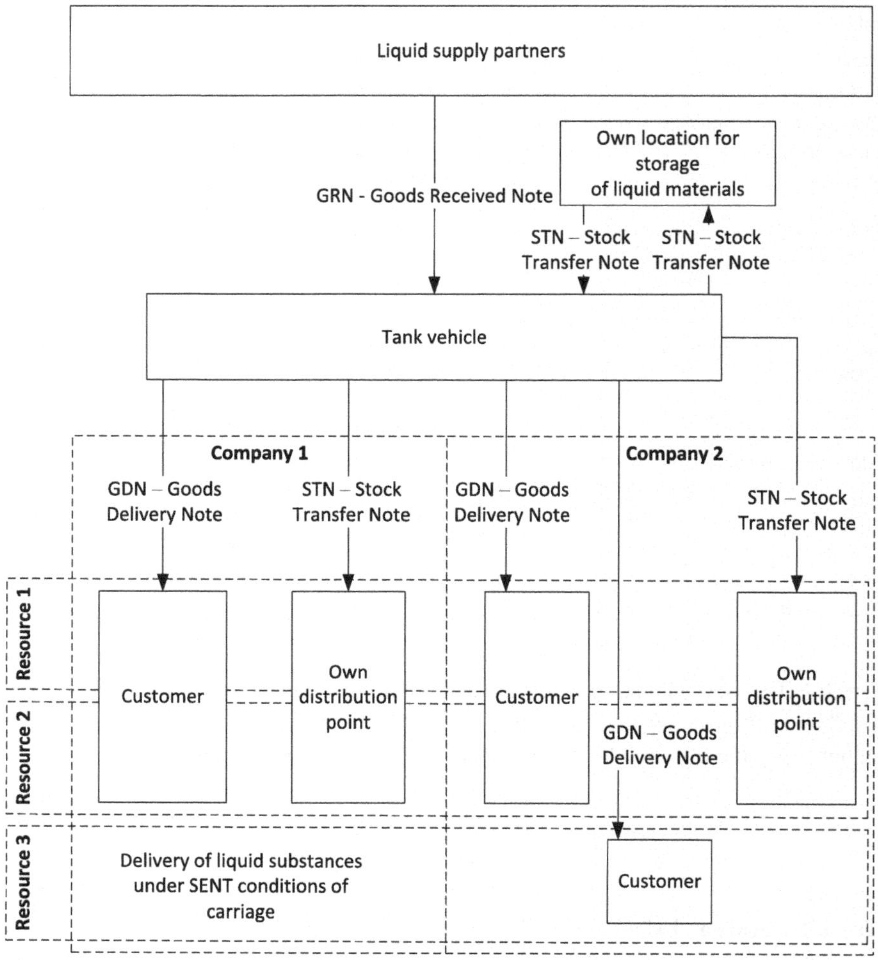

Fig. 1 Liquid products flow diagram supplemented by corresponding types of documentation, own study

- The delivery point, including location, time window, type and quantity of liquid products ordered, exact unloading time, and vehicle availability.
- Loading space capacity, permitted driving times and required by law rest periods, gross vehicle weight limits, direct and indirect costs.
- Road data of the research area, average road speeds, road restrictions.

This set of essential data elements is consistent with the supply chain elements identified during the literature review, which are presented above in Table 1.

4.2 *Supply Chain Analysis Iteration Structure and Key Parameters*

In relation to the parameters included in the analysis, a selection of parameters that are subject to change from the perspective of the analysed company was made together with representatives of a company in the liquid transport industry. Therefore, the key parameters of the company were identified and their impact on the efficiency of the transport processes had to be checked. Due to the best tool functionality, the decision was made to use the ArcGIS Pro analysis tool with the Network Analyst add-on, which enables digital simulation of transport process efficiency. A series of six consecutive iterations were developed. The number and range of iterations were proposed based on the parameters identified in Table 1. On the basis of the literature review, and the expert panel, the parameters to be modified in subsequent iterations were selected. For each iteration, a set of control parameters subject to change or exclusion was adopted. The first of these (0-0) reflected the current as-is logistics model. It was decided to take this step to validate the computational model created in the tool. The accuracy of the actual model was assessed by comparing the total distance travelled by the company's vehicles with the number of litres of liquid products delivered. This proved that the created computational model reflected the distance parameter of the transport processes and covered customer demand in total. As a result, it ensured that modifications in subsequent iterations did not affect the level of customer service and that the actual demand–supply model was replicated in the simulation environment. Each subsequent alternative distribution scenario was designed to verify the impact of selected supply chain parameters on overall efficiency. In order to clearly present all parameters subject to change in subsequent iterations Table 2 was created. It contains a summary of the scenarios and parameters subject to change in each iteration. The table includes information on all iterations and parameters that were verified in subsequent simulations in the ArcGIS Network tool. The simulations conducted within iterations are based on data from a two-week transport operational period. Each scenario contains only one iteration. The term scenario has been introduced to make it more convenient to identify a specific iteration and to indicate the order of the simulations. Scenario 0 is reserved for the as-is iteration without any changes. Subsequent iterations take into account changes in key parameters and have been performed in numerical order.

In the 0-0 iteration, the model calculated the target number of routes and the distance travelled by the 19,000 and 32,000 L tankers. It was determined that the primary model, which reflected the performance of the conditions of the real-life model, accurately calculating the number of routes and distances travelled in the simulation environment. The efficiency ratio calculated for the 0-0 iteration in litres per kilometre represents a reference value for subsequent comparisons of efficiency levels in each iteration.

The research carried out in iterations 4th and 5th was based on a hypothetical regionalised logistic service area. This was referred to the assumption that the area covered would be divided into relatively equal parts, based on the most favourable

Table 2 Control parameters considered in each iteration

	Iteration no./iteration code identification					
	0	1st	2nd	3rd	4th	5th
	0-0	0-1	1-0	1-0-1	1-1	1-1-1
Main control parameter	*CM*	*TMS*	*TMS + COM*	*TMS + COM + AVA*	*TMS + COM + REG*	*TMS + COM + AVA + REG*
Order of delivery drop off points along the route according to historical data	+	–	–	–	–	–
Order of delivery drop off points according to the optimisation algorithm	–	+	+	+	+	+
Separation of Company 1 and Company 2 distribution into separate tank vehicles	+	–	–	–	–	–
A consolidation of Company 1 and Company 2 distribution	–	+	+	+	+	+
Unloading time according to historical data	+	–	–	–	–	–
Unloading time calculated as = 1 min * 300 L + 5 min	–	+	+	+	+	+
Delivery days according to historical data. (As per customer request)	+	+	+	+	+	+
Driving time according to historical data	+	–	–	–	–	–
Driving time 9 h maximum	–	+	+	+	+	+

(continued)

Table 2 (continued)

	Iteration no./iteration code identification					
	0	1st	2nd	3rd	4th	5th
	0-0	0-1	1-0	1-0-1	1-1	1-1-1
Maximum payload according to historical data	+	–	–	–	–	–
Maximum load capacity 32 000/19 000	–	+	+	+	+	+
Transport arranged by 19 000 L tank vehicles only	–	+	+	+	+	+
Start and finish at the base depot	–	+	+	+	+	+
No regionalization	+	+	+	+	-	-
Split of customers into two distribution areas—North domestic district and South domestic district	–	–	–	–	+	+
Liquid supply partners could only sell to customers with signed contract	+	+	+	–	+	–
Liquid supply partners can issue all types of liquid products	–	–	–	+	–	+

Source own elaboration
All iteration assumption presented within the Table 2
CM Current model
TMS the adoption of an optimisation algorithm relating to the route sequence parameter, distance and driving time
COM A combination of distribution of products belonging to company C1 and C2
AVA Availability of all products at all loading points
REG Regionalisation of distribution

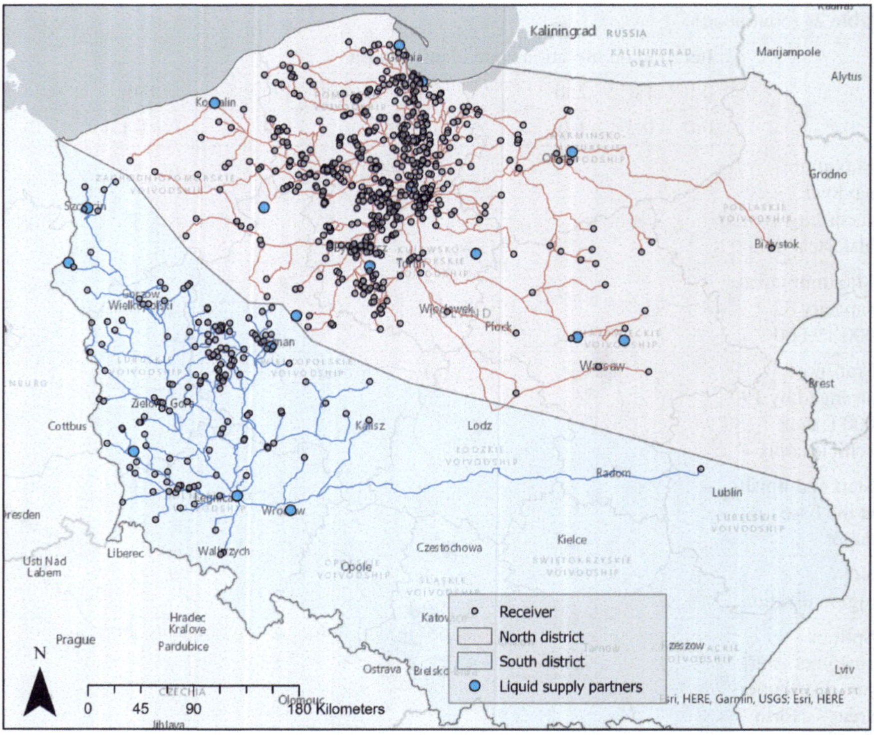

Fig. 2 Distribution districts North and South

travel time from the supplier to the customer. A drawing of the two main regions and an example of the route planning results of one of the iterations are shown below in Fig. 2. Regionalisation has been introduced in iterations 1-1 and 1-1-1.

All six iterations (including the 0-0 iteration) were executed within the same simulation GIS-class tool, according to the control parameters presented in Table 2. The results of the simulations are outlined in the next section.

5 Results and Discussion

The result of the simulations carried out shows a constant number of litres handled in the subsequent iterations (Table 3). The same quantity was handled in each iteration, which fully covers the demand expressed in the as-is model. Each of the subsequent iterations, taking into account the changes in the key parameters of the as-is logistics model, shows a decrease in the number of kilometres. However, in subsequent iterations the number of routes committed to meeting demand has increased. In particular, there has been a significant increase in the use of 32,000 L tankers. A much smaller increase in the number of 19,000 L tankers was observed.

The results obtained in 5th iteration indicate a positive impact of applying distribution regionalisation which is consistent with the results of the Khalfan et al. (2014) study. However, it should be also noticed that a positive effect is obtained in the case of a change in functionality and availability of products of different types at tanker

Table 3 Efficiency changes in subsequent iterations

Iteration no	0 (as-is)	1st	2nd	3rd	4th	5th
Iteration code identification	0-0	0-1	1-0	1-0-1	1-1	1-1-1
Main control parameter change	CM	TMS	TMS + COM	TMS + COM + AVA	TMS + CCM + REG	TMS + COM + AVA + REG
Number of kilometres travelled	58 284	56 711	56 415	47 630	56 360	47 832
Total number of litres of liquid products	6 332 484	6 332 484	6 332 484	6 332 484	6 332 484	6 332 484
Liquid product no. 1	5 453 674	5 453 674	5 453 674	5 453 674	5 453 674	5 453 674
Liquid product no. 2	600 731	600 731	600 731	600 731	600 731	600 731
Liquid product no. 3	147 008	147 008	147 008	147 008	147 008	147 008
Liquid product no. 4	6 013	6 013	6 013	6 013	6 013	6 013
Liquid product no. 5	125 058	125 058	125 058	125 058	125 058	125 058
Total number of routes	186	281	275	237	280	233
Number of vehicle routes (Tank capacity 32 000 L)	166	260	260	217	251	211
Number of vehicle routes (Tank capacity 19 000 L)	20	21	15	20	29	22
Overall Iteration Efficiency (litre/kilometres)	108.65	111.66	112.25	132.95	112.36	132.39
Change in efficiency level relative to the current model (As-Is)		2.77%	3.31%	22.37%	3.41%	21.85%

Source own elaboration
All iteration assumption presented within the Table 3
CM Current model
TMS the adoption of an optimisation algorithm relating to the route sequence parameter, distance and driving time
COM A combination of distribution of products belonging to company C1 and C2
AVA Availability of all products at all loading points
REG Regionalisation of distribution

loading locations. This indicates a similar effect on efficiency as that indicated in the Garcia and You (2015) study.

The decrease in efficiency was not observed in any of the iterations. This may indicate that the model currently in-use is inefficient. It confirms the assumption that models where management methods have not been implemented and not supported by appropriate TMS tools may show variable efficiency. This is usually due to the manual execution of a process with variable standards (Džubáková and Kopták 2017).

6 Conclusions

The results were used to estimate the savings potential (in the local currency, the Polish zloty PLN). Assuming an average rate of PLN 4.8 for driving 1 km by tanker truck, the potential for annual savings was identified as between PLN 171,771 and PLN 1,163,416 (about EUR 40,300 - EUR 272,900), according to the selected variant of supply chain set-up. Detailed information is presented in Table 4.

Table 4 Potential for savings per iteration

Iteration no	0 (AS-IS)	1st	2nd	3rd	4th	5th
Main control parameter change	CM	TMS	TMS + COM	TMS + COM + AVA	TMS + COM + REG	TMS + COM + AVA + REG
Sum of kilometres travelled in analysed period	58 284	56 711	56 415	47 630	56 360	47 832
Estimated amount of km per year	1 515 384	1 474 486	1 466 790	1 238 380	1 465 360	1 243 632
Distance reduction in each iteration [year]		40 898	48 594	277 004	50 024	271 752
Reduction of kilometres in compare with current iteration [%]		2.77%	3.31%	22.37%	3.41%	21.85%
Estimated savings per iteration [PLN—Polish Zloty]		PLN 171,771	PLN 204 094	PLN 1 163 416	PLN 210,100	PLN 1 141 358

Source own elaboration
All iteration assumption presented within the Table 3
CM Current model
TMS the adoption of an optimisation algorithm relating to the route sequence parameter, distance and driving time
COM A combination of distribution of products belonging to company C1 and C2
AVA Availability of all products at all loading points
REG Regionalisation of distribution

Managerial Implications

Increasing the loading capacity of different liquid raw materials in multiple locations has a significant impact on operational efficiency. A study using simulations has shown how decisions that increase the availability of raw materials in different regions can influence route optimisation. An example of this is the unification of raw material availability in disparate geographical areas, which increases flexibility in planning. The simulation also included the possibility of stationing vehicles at liquid raw material supply partners, which led to even better results in terms of efficiency and logistics cost reduction.

Based on the simulations, it was shown that a positive impact on distribution efficiency can be achieved by merging the deliveries of Company 1 and Company 2 (efficiency increase of 3.31%). At the same time, the use of advanced route modelling algorithms can influence an overall efficiency increase of 2.27%. Allowing the loading of liquid products of different types at all loading points leads to efficiency gains between 21.85 and 22.37%. The introduction of regionalisation shows a variable impact on efficiency and shows the best influence in combination with other parameter changes, such as product availability. The introduction of regionalisation alone can increase efficiency by 3.41%.

Practical implications

The results of the study indicate that by increasing commodity availability and modelling the functionality of global supply chain participants, process efficiency can be significantly improved. By combining the extension of functionality with the regionalisation of transport services, comparable results can be achieved. It is important to select the right set of parameters to be changed in the context of the supply chain under analysis in order to have an optimal balance between cost and impact.

Further research and limitations

Further research could be carried out to measure efficiency in the form of iterative simulations using different modes of transport, including intermodal freight. Both the research method used and the form of the results suggest that the proposed method of indicating the most efficient supply chain configuration may be sufficient and scalable. Limitations encountered during the research indicate that the complexity of current supply chains results in vast datasets, forcing detailed analyses to be performed using the ArcGis tool with the Network add-on only for a limited timeframe. Furthermore, the limitation of this study results from the characteristics of the analysed case study. Further research will test the proposed approach also for different case studies in different industries.

Acknowledgements Damian Dubisz would like to thank the Ministry of Education and Science for funding his research within the framework of the Applied Doctorate Program of the Ministry of Education and Science of Poland (Program Doktorat Wdrozeniowy Ministerstwa Edukacji i Nauki) implemented in years 2021-2025 (Contract No. DWD/5/0015/2021 dated 23/12/2021).

References

Abdulla MF, Musa H (2021) Mediation model of logistics service supply chain (LSSC) factors affecting organisational performance. Int J Sustain Constr Eng Technol 12(5). https://doi.org/10.30880/ijscet.2021.12.05.030

Acero R, Torralba M, Pérez-Moya R, Pozo JA (2019) Order processing improvement in military logistics by Value Stream Analysis lean methodology. Procedia Manuf 41:74–81. https://doi.org/10.1016/j.promfg.2019.07.031

Almalki A, Gokaraju B, Mehta N, Doss DA (2021) Geospatial and machine learning regression techniques for analyzing food access impact on health issues in sustainable communities. ISPRS Int J Geo-Inf 10(11):745. https://doi.org/10.3390/ijgi10110745

Arena M, Azzone G, Grecchi M, Piantoni G (2021) How can the waste management sector contribute to overcoming barriers to the circular economy? Sustain Dev 29(6):1062–1071. https://doi.org/10.1002/sd.2202

Bruinen De Bruin Y, Hakkinen P, Lahaniatis M, Papameletiou D, Del Pozo C, Reina V, Van Engelen J, Heinemeyer G, Viso AC, Rodriguez C, Jantunen M (2007) Risk management measures for chemicals in consumer products: documentation, assessment, and communication across the supply chain. J Expo Sci Environ Epidemiol 17(S1):S55–S66. https://doi.org/10.1038/sj.jes.7500587

Cadarso MÁ, Gómez N, López LA, Tobarra MÁ (2016) Calculating tourism's carbon footprint: measuring the impact of investments. J Clean Prod 111:529–537. https://doi.org/10.1016/j.jclepro.2014.09.019

Cai J, Liu X, Xiao Z, Liu J (2009) Improving supply chain performance management: a systematic approach to analyzing iterative KPI accomplishment. Decis Support Syst 46(2):512–521. https://doi.org/10.1016/j.dss.2008.09.004

Cambero C, Sowlati T (2014) Assessment and optimization of forest biomass supply chains from economic, social and environmental perspectives—a review of literature. Renew Sustain Energy Rev 36:62–73. https://doi.org/10.1016/j.rser.2014.04.041

Chen S, Leng X, Luo K (2024) Supply chain digitalization and corporate ESG performance. Am J Econ Sociol 83(4):855–881. https://doi.org/10.1111/ajes.12596

Cichosz M, Pluta-Zaremba A (2019) How to improve freight transport emissions' management? Logforum 15(1):93–105. https://doi.org/10.17270/J.LOG.2019.312

Dubisz D, Golińska-Dawson P, Koliński A (2022a) Measuring CO_2 emissions level for more sustainable distribution in a supply chain. Eng Appl Sci Res 49:804810. https://doi.org/10.14456/EASR.2022.78

Dubisz D, Golinska-Dawson P, Zawodny P (2022b) Measuring CO_2 emissions in e-commerce deliveries: from empirical studies to a new calculation approach. Sustainability 14(23):16085. https://doi.org/10.3390/su142316085

Durugbo CM, Al-Balushi Z (2023) Supply chain management in times of crisis: a systematic review. Manag Rev Q 73(3):1179–1235. https://doi.org/10.1007/s11301-022-00272-x

Džubáková M, Kopták M (2017) Work standardisation in logistics processes. Qual Innov Prosper 21(2):109–123. https://doi.org/10.12776/qip.v21i2.880

Egri P, Dávid B, Kis T, Krész M (2021) Robust facility location in reverse logistics. Ann Oper Res. https://doi.org/10.1007/s10479-021-04405-5

Emami P, Marzban A (2023) The synergy of artificial intelligence (AI) and geographic information systems (GIS) for enhanced disaster management: opportunities and challenges. Disaster Med Public Health Prep 17:e507. https://doi.org/10.1017/dmp.2023.174

Gao J, Hua G, Huo B (2024) Green finance policies, financing constraints and corporate ESG performance: insights from supply chain management. Oper Manag Res. https://doi.org/10.1007/s12063-024-00509-w

Gao R, Zhou K, Zhou Y, Yang C (2020) Research on the fluid characteristics of cemented backfill pipeline transportation of mineral processing tailings. Alex Eng J 59(6):4409–4426. https://doi.org/10.1016/j.aej.2020.07.047

Garcia DJ, You F (2015) Supply chain design and optimization: challenges and opportunities. Comput Chem Eng 81:153–170. https://doi.org/10.1016/j.compchemeng.2015.03.015
Gelman A, Rubin DB (1992) Inference from iterative simulation using multiple sequences. Stat Sci 7(4). https://doi.org/10.1214/ss/1177011136
Ghosh A, Fedorowicz J (2008) The role of trust in supply chain governance. Bus Process Manag J 14(4):453–470. https://doi.org/10.1108/14637150810888019
Gligor DM, Holcomb MC (2012) Understanding the role of logistics capabilities in achieving supply chain agility: a systematic literature review. Supply Chain Manag Int J 17(4):438–453. https://doi.org/10.1108/13598541211246594
Golinska-Dawson P (2020) Towards circular economy transition—developing the innovative sustainable practices in logistics industry. In: Golinska-Dawson P (ed) Logistics operations and management for recycling and reuse. Springer, Berlin Heidelberg, Berlin, Heidelberg, pp 3–18
Goliński M, Szafrański M (2019) Application of corporate social responsibility for competency management—case study. In: Golinska-Dawson P, Spychała M (eds) Corporate social responsibility in the manufacturing and services sectors. Springer, Berlin Heidelberg, Berlin, Heidelberg, pp 3–18
Kannegiesser M, Günther H-O (2014) Sustainable development of global supply chains—part 1: sustainability optimization framework. Flex Serv Manuf J 26(1–2):24–47. https://doi.org/10.1007/s10696-013-9176-5
Kara S, Rugrungruang F, Kaebernick H (2007) Simulation modelling of reverse logistics networks. Int J Prod Econ 106(1):61–69. https://doi.org/10.1016/j.ijpe.2006.04.009
Khalfan MMA, Maqsood T, Noor MA (2014) Relationships among supply chain participants: the case of Australia and Malaysia. Int J Procure Manag 7(4):376. https://doi.org/10.1504/IJPM.2014.063167
Kolinski A, Horzela A, Cudzilo M, Domanski R (2020) Reference model of information flow in business relations with 4PL operator. In: Kolinski A, Dujak D, Golinska-Dawson P (eds) Integration of information flow for greening supply chain management. Springer International Publishing, Cham, pp 19–45
Li D, O'Brien C (2001) A quantitative analysis of relationships between product types and supply chain strategies. Int J Prod Econ 73(1):29–39. https://doi.org/10.1016/S0925-5273(01)00094-9
López C, Ruiz-Benítez R (2020) Multilayer analysis of supply chain strategies' impact on sustainability. J Purch Supply Manag 26(2):100535. https://doi.org/10.1016/j.pursup.2019.04.003
Martyushev NV, Malozyomov BV, Khalikov IH, Kukartsev VA, Kukartsev VV, Tynchenko VS, Tynchenko YA, Qi M (2023) Review of methods for improving the energy efficiency of electrified ground transport by optimizing battery consumption. Energies 16(2):729. https://doi.org/10.3390/en16020729
Mogale DG, De A, Ghadge A, Aktas E (2022) Multi-objective modelling of sustainable closed-loop supply chain network with price-sensitive demand and consumer's incentives. Comput Ind Eng 168:108105. https://doi.org/10.1016/j.cie.2022.108105
Rakhimov O, Eshev S, Rakhmatov M, Saidov I, Boymurodov F, Rayimova I (2021) Improved pump for transporting liquid feed mixtures through pipes on farms. E3S Web Conf 263:04046. https://doi.org/10.1051/e3sconf/202126304046
Sayan Z, Sürücü L (2024) The role of transformational leadership and employee productivity in logistics performance. Logforum 20(1):1–10. https://doi.org/10.17270/J.LOG.000965
Singh RK, Modgil S, Shore A (2024) Building artificial intelligence enabled resilient supply chain: a multi-method approach. J Enterp Inf Manag 37(2):414–436. https://doi.org/10.1108/JEIM-09-2022-0326
Stachowiak A, Hadaś Ł, Cyplik P, Fertsch M (2013) Decision model for sustainable and agile resources management. IFAC Proc 46(9):1140–1145. https://doi.org/10.3182/20130619-3-RU-3018.00483

Tao Z, Zheng Q, Kong H (2018) A modified gravity p-median model for optimizing facility locations. 6:421–434. https://doi.org/10.21078/JSSI-2018-421-14

Werner-Lewandowska K, Golińska-Dawson P, Cyplik P (2024) Stimulators of transition toward circular supply chain—exploratory studies in electrical and electronic equipment industry. Logforum 20(1):83–96. https://doi.org/10.17270/J.LOG.000973

Zając J, Dubisz D, Koliński A (2024) Multi-criteria analysis of road transport service providers as a method to support decision-making within sustainable supply chains. Sci J Marit Univ Szczec 151(79). https://doi.org/10.17402/614

Stochactic Programming in Supply Chain Optimalization

Drago Pupavac, Ljudevit Krpan, and Ivica Baković

Abstract The main goal of this chapter is to investigate the possibilities of optimizing the supply chain delivery service. For creating an optimal delivery service we used the method of stochastic programming. The method of stochastic programming was chosen for the reason that it allows the decision maker to explicitly analyze uncertainties and control risk. The chapter presents a comparative presentation of the optimization of delivery services using the linear programming method and the stochastic programming method. For this purpose, a computer-supported model was developed in an MS Excel spreadsheet. The main finding of this chapter points to the conclusion that stochastic programming minimizes costs by taking into account all possible scenarios in a way to achieve a robust solution. Despite their advantages and the growing and numerous uncertainties in supply chain operations, stochastic models are still not widely accepted in business practice.

Keywords Supply chains · Stochastic programming · MS Excel spreadsheet · Optimization

1 Introduction

For many current supply chain issues, the data cannot be fully known for a number of reasons. This is primarily due to the fact that many data refer to future events (what will be the demand for the product, what will be the prices of the product in the future period) and it is not possible to predict them with certainty. The research

D. Pupavac (✉)
Polytechnic of Rijeka, Rijeka, Croatia
e-mail: drago.pupavac@veleri.hr

L. Krpan
University of North Croatia, Varaždin, Croatia
e-mail: ljudevit.krpan@unin.hr

I. Baković
University of Applied Sciences Hrvatsko Zagorje Krapina, Krapina, Croatia
e-mail: ivica.bakovic@vhzk.hr

A. Kolinski and M. Adamczak (eds.), *Digitalisation of the Greening Supply Chain*, EcoProduction, https://doi.org/10.1007/978-3-031-88918-9_6

question to which the answer is sought in this paper is, how can supply chain management improve the decision-making process in conditions of increasing uncertainty? Working hypothesis of this paper is: Decisions that do not take uncertainties into account can cause incalculable costs in supply chain operations and loss of end users. Accordingly, the research subject of importance for this paper is the role of stochastic programming in making optimal decisions in conditions of unreliable data. The importance of stochastic programming becomes especially important at a time when there is uncertainty not only about supply, markets, prices, demand, but also about sources of supply. Uncertainties are more than obvious as downstream and upstream in supply chains. Research results are based on the analysis of a practical example of LPG distribution using mathematical linear programming and stochastic programming methods. For this purpose, a computer-supported model was developed that can serve supply chain managers in finding the best business decisions. The main finding of this paper refers to the limitations of linear programming due to its deterministic treatment of the future, that is, to the advantages of the stochastic programming method in relation to the linear programming method when making decisions under the conditions of the occurrence of future events.

2 Literature Review

Supply chains are essential for the transportation and distribution of goods from suppliers to customers (Zelenika and Pupavac 2008). Supply chains with complex structure are more vulnerable to disruptions in the new business arena. Therefore, mathematical linear programming models that treat the future deterministically are less and less applicable. An extension of linear and mixed integer programming, called stochastic programming solves this problem. The basic concept is to consider several uncertain future scenarios at the same time, each with a different probability of happening. A popular concept in stochastic programming is the two-stage approach. A logical approach to solving such problems requires that one decision be made immediately, then wait for some uncertainty to be resolved, and then make another decision based on what happened. This paradigm is called the recourse model in the literature (Beale et al. 1986; Hashnayne 2019).

$$\begin{aligned} &\text{Min} \quad c^T x + EwQ(x,w) \\ &\text{Subject to} \quad Ax = b \\ &\qquad x \geq 0 \\ &\text{Min} \quad d(w)^T y \\ &\text{Subject to} \quad T(w)x + W(w)y(w) = h(w) \end{aligned} \tag{1}$$

The first linear program minimizes the direct costs of the first period, cTx, plus the expected corrective costs, Q(x,w), for all possible scenarios while satisfying the constraint of the first period, Ax = b.

Operational models of problems in supply chains offer a ripe set of applications for stochastic programming (Powell and Topaloglu 2003). Zhuang et al. (2023) proposed two-stage stochastic programming model to optimizes the new-line-opening and rerouting decisions in the face of uncertain transport demands and transportation times. Manotas Niño and Gómez Vizcaíno (2010) proposed stochastic programming for designing a good logistic network. Owen and Daskin (1998) research possibilities of stochastic programming in the facility location component of supply chain network design under uncertainty. Hassani et al. (2000) considered a two-stage model for multi-period capacity planning of supply chain networks. Shapiro (2001) demonstrated stochastic programming in demand forecasts for supply chain products. Gutierrez et al. (1996) used stochastic programming for seeks network configurations that are good (nearly optimal) for a variety of scenarios of the design variables. Nezhadroshan et al. (2020) used stochastic programming to define new transportation network's routes after an earthquake to provide a plan against uncertainty in whole levels of supply chain along with its availability. Mahtab et al. (2021) used stochastic programming to develop logistics model to assist disaster management officials in making optimal pre- and post-disaster decisions. Dillon et al. (2017) propose a two-stage stochastic programming model for defining optimal periodic review policies for red blood cells inventory management.

3 Practical Example

In order to clearly present the possibility of optimizing the delivery service using stochastic programming, we will start from the following example. The assumption is that the questionable chain consists of a supplier of liquefied petroleum gas (LPG) located in Hungary, a distributor of liquefied petroleum gas located in northern Croatia and customers of liquefied petroleum gas in Croatia (cf. Fig. 1).

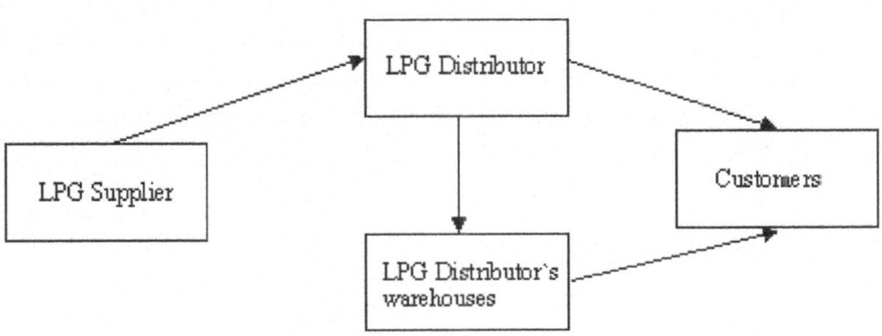

Fig. 1 LPG supply chain, own study

Table 1 Three possible scenarios next winter

Scenario	Probability	LPG Cost (€/kg)	Demand (000 kg)
Normal	1/3	0,5	2000
Cold	1/3	0,6	3000
Very cold	1/3	0,75	3600

Source own study

When an LPG distributor buys LPG, it usually delivers one part directly to customers and stores the other part. When it sells LPG, it delivers it either from storage or from new orders. Accordingly, the management of the distribution company should make the following decisions: (1) how much gas to purchase and deliver, (2) how much gas to purchase and store, (3) how much gas to take from storage and deliver to customers. Management decisions will depend on current and future gas prices, storage costs, storage space size, and demand in each period.

The assumption is that consumers are facing an average cold winter and that the next winter can have one of three scenarios: an average cold winter (normal Winter—NW), an above average cold winter (Cold Winter—CW) or a very, very cold winter (VCW). To formulate this problem as a stochastic linear program, we must first determine the uncertainty in the model. The most frequently used method is to formulate scenarios, and to associate a certain probability of occurrence with each scenario. Each of the possible scenarios has different data as shown in the Table 1.

LPG demand and LPG prices are rising as the weather gets colder. Storage costs are constant, for one unit of gas €0.1 per year.

Management doesn't know which of these three scenarios will happen next year, but they want current procurement decisions to be as good as possible to minimize expected costs. It should be kept in mind that at the moment when they make the second procurement decision, they will know exactly which of the three scenarios actually happened.

Formulating a stochastic linear program aims to minimize the costs of the first period plus the expected costs of the second period. The dynamic programming method solves multistage stochastic programming problems in an efficient way. Stochastic programming requires a lot of data and calculations, which is why managers often decide to use simpler solutions when making business decisions. Therefore, in this scientific discussion, a computer-supported model will be developed in the MS Excel spreadsheet, which should remove this deficiency and enable managers to use it more easily and make decisions close to the optimum.

4 Mathematical Model

Starting from model (1) and a practical example, the set problem can be presented with deterministic constraints by introducing different variables y for the second period, and for each scenario separately.

This formulation is called the deterministic equivalent.

$$\begin{aligned} \text{Min} \quad & c^T x + \sum_{i=1}^{N} p_i d_i^T y_i \\ \text{Constrains} \quad & Ax = b \\ & T_i x + W_i y_i = h_i, \ i = 1, \ldots, N \\ & x \geq 0 \\ & y_i \geq 0 \end{aligned} \tag{2}$$

where N is the number of scenarios and pi is the probability that a certain scenario will occur.

For the previously posed problem with three possible scenarios we have

$$\begin{aligned} \text{Min} \quad & c^T x + p_1 d_1^T y_1 + p_2 d_2^T y_2 + p_3 d_3^T y_3 \\ \text{Constrains} \quad & Ax = b \\ & T_1 x + W_1 y_1 = h_1 \\ & T_2 x + W_2 y_2 = h_2 \\ & T_3 x + W_3 y_3 = h_3 \\ & x,\ y_i \geq 0 \end{aligned} \tag{3}$$

Here we have only one decision for the first period, x, and N decisions in the second period, one for each scenario. The decision in the first period cannot anticipate all three scenarios and must be solvable for each scenario. Therefore, Ax = b and Ti x + Wi yi = hi for i = 1,…,N. So we simultaneously solve x and y for all decisions, choosing x to be somehow optimal for all three scenarios.

5 Research Results and Discussion

The costs for each scenario are given below (Tables 2, 3 and 4):

We don't know which of these three scenarios will happen next year, but we want our current procurement decisions to be as good as possible to minimize expected costs.

Table 2 Normal winter—normal winter

Year	Purchase to use (000)	Purchase to Store (000)	Storage (000)	Cost (000 €)
1	2000	0	0	1000
2	2000	0	0	1000

Total Cost = 2000 €
Source own study

Table 3 Normal winter—cold winter

Year	Purchase to use (000)	Purchase to store (000)	Storage (000)	Cost (000 €)
1	2000	0	0	1000
2	3000	0	0	1800

Total Cost = 2800 €
Source own study

Table 4 Normal winter—very cold winter

Year	Purchase to use (000)	Purchase to store (000)	Storage (000)	Cost (000 €)
1	2000	3600	3600	3160
2	0	0	0	0

Total Cost = 3160 €
Source own study

Below is a presentation (comparison) of problem solutions using the linear programming method and the stochastic programming method. When solving a problem using the linear programming method, all random variables will be replaced with their expected values. For the purposes of this approach, it seems appropriate to start from average values (cf. Table 5).

The solution to this problem will be approached using an MS Excel spreadsheet. After the data on procurement and storage costs and data on mandatory procurement and demand have been entered into the table, it is necessary to enter the formulas in the corresponding addresses. For example, the formula = (SUM(C9:E9)/3) was entered in the address field C7, which determined the average demand for all three scenarios. The closing stocks at the end of the first and at the end of the second year were also determined by that formula. The closing stocks from the first period C10 represent the starting stocks for each of the three possible scenarios, thus establishing a link between the decision in the first and second year. Procurement costs are determined by the formula = B6 * B3, which is pasted into the address field B12, and then copied into the address string C12:E12. Inventory costs were determined in a similar way. The optimal solution to the problem of distribution of liquefied petroleum gas using the linear programming method is given in the address field C15 and determined by the formula = B14 + C14. The costs of corrective actions are located in the address string C14:E14. The expected costs are given in the address field F15 and are determined by the sum of the costs for each of the three possible scenarios and

Table 5 Average expected values

Year	Purchase cost LPG (€/kg)	Demand for LPG (000 kg)
1	0.5	2000
2	0.617	2866.67

Source own study

	A	B	C	D	E	F
2	Unit Cost	NW	NW	CW	VCW	
3	Purchase	0,50	0,50	0,60	0,75	
4	Storage	0,10	0,10	0,10	0,10	
5	Obligatory Purchase	200	200	200	200	
6	**Optimal Purchase**	**2000**	**2866,666667**	**133,3333333**	**733,3333333**	
7	Max Purchase	2000	2866,666667	133,3333333	733,3333333	
8	Beginning inventory	0	-	-	-	
9	Demand	2000	2000	3000	3600	
10	Ending inventory	-	2866,67	-	-	
11	Ending invetory II		866,67			
12	Purchase cost	1000,00	1433,33	80,00	550,00	
13	Storage cost	0,00	286,67	0,00	0,00	
14		1000,00	1720,00			
15	Optimal solution		2720,00			
16	Corrective cost		86,67	80,00	550,00	
17	Scenario Cost		2806,67	2800,00	3270,00	
18	Probability		0,33	0,33	0,33	
19	Expected Cost		935,55	933,32	1089,99	2958,86

Fig. 2 Computer supported model for solving linear programming model, own study

the probability of their occurrence. Accordingly, in the address field F19 there is a formula = SUM(C19:E19) which simultaneously represents the objective function.

The card for solving the mentioned problem has the following layout (Fig. 2):

Set Taregt Cell = F19 min.

Variables B6:E6.

Constrains.

B6:E6 $\geq$ 0.

B6:E6 $\geq$ B7:E7.

The expected total costs for all three scenarios amount to

$\frac{1}{3}$ **2 806,67** + $\frac{1}{3}$ **2 800** + $\frac{1}{3}$ **3 270** = **2,958.86 €**.

By solving the set problem using the method of stochastic linear programming, the solution given by Fig. 3 is obtained.

The expected total costs for all three scenarios amount to.

$\frac{1}{3}$ **2 200** + $\frac{1}{3}$ **2 800** + $\frac{1}{3}$ **3 400** = **2,799.97 €**.

Based on the obtained solutions, several conclusions can be drawn. First, knowing the individual optimal solutions, it is evident that the managers of the gas distribution company should decide how much gas to store. As solutions, the possibilities to store nothing or to store an amount of 3,600 tons of LPG stand out. The solution obtained on the basis of average data offers a storage option of 2,866.67 tons of LPG. The solution of the mentioned problem using the method of stochastic programming offers as a solution the storage of LPG stocks in the amount of 2,000 tons. The obtained solution does not correspond to any previously obtained optimal solution, but it has built-in a high level of flexibility in order to best meet the uncertain demand in the second year. Second, stochastic programming prevents management from "gambling" with

	A	B	C	D	E	F	G
2		Unit Cost	NW	NW	CW	VCW	
3		Purchase	0,50	0,50	0,60	0,75	
4		Storage	0,10	0,10	0,10	0,10	
5		Obligatory Purchase	200	200	200	200	
6		**Optimal Purchase**	**2000**	**2000**	**1000**	**1600**	
7		Max Purchase	2000	2000	1000	1600	
8		Beginning inventory	0	-	-	-	
9		Demand	2000	2000	3000	3600	
10		Ending inventory	-	2.000	-	-	
11		Ending invetory II					
12		Purchase cost	1000,00	1000,00	600,00	1200,00	
13		Storage cost	0,00	200,00	0,00	0,00	
14			1000,00	1200,00			
15		Optimal solution		2200,00			
16		Corrective cost		0	600,00	1200,00	
17		Scenario Cost		2200,00	2800,00	3400,00	
18		Probability		0,33333	0,33333	0,33333	
19		Expected Cost		733,326	933,324	1133,322	**2799,972**

Fig. 3 Computer supported model for solving stochastic linear programming model, own study

decision-making, because such an approach could be disastrous for both the gas distributor and its customers. For example, if the management were convinced that the next winter would be very, very cold and decided to store 3,600 tons of LPG, and the next winter would be moderately cold, this would result in total costs for the gas distributor being 66% higher than the optimal solution, or 55.91% higher compared to the solution obtained by the method of stochastic linear programming. Third, the set model minimizes the expected costs for each of the possible scenarios, and its optimal solution gives costs of €2,200, €2,800 and €3,400 in each scenario. Individually observed, some of the expected costs could be unacceptable for the management of the company, for example €3,400, but by determining the probability of occurrence of each of the outcomes, a more objective picture is obtained. Fourth, the difference in expected costs in the amount of 158.89 thousand euros is an advantage of stochastic programming compared to the solution obtained based on expected average values. Fifth, the obtained costs per stochastic solution are greater than or equal to the optimal solutions for each individual scenario (2,200 $\geq$ 2,000, 2,800 $\geq$ 2,800, and 3,400 $\geq$ 3,160). Since it is almost impossible to determine future events with certainty, the stochastic solution sacrifices the minimum costs of individual scenarios in search of a robust solution for all three potentially possible scenarios.

6 Conclusions and Further Research

The developed model simultaneously determines an optimal contingency plan for each scenario and an optimal here-and-now plan that optimally hedges against these contingency plans. Optimization entails minimization of expected costs, where expected costs refer to multiplying costs with each scenario by its probability of occurrence. The solutions of the practical example obtained by the stochastic programming method fully support the hypothesis that decisions that do not consider the uncertainty of the result can have extremely high costs and loss of end users. Stochastic programming promotes a scientifically based approach to making business decisions within supply chains under conditions of contingency. Flexibility and robustness are the main features of the obtained stochastic programming solutions and decisions based on them. The main shortcoming of this work stems from the fact that the developed computer supported model includes only two periods. So, in the future research we will extend the model for an arbitrary number of periods.

References

Beale EML, Dantzig GB, Watson RD (1986) A first order approach to a class of multi-time-period stochastic programming problems. In: Prékopa A, Wets RJB (eds) Stochastic programming 84 Part I. Mathematical programming studies, vol 27. Springer, Berlin, Heidelberg

Dillon M, Oliveira F, Abbasi B (2017) A two-stage stochastic programming model for inventory management in the blood supply chain. Int J Prod Econ 187(C):27–41

Gutierrez GJ, Kouvelis P, Kurawala AA (1996) A robustness approach to uncapacitated network design problems. Eur J Oper Res 94:362–376

Hashnayne A (2019) Formulation of two-stage stochastic programming with fixed recourse. Br Int Exact Sci (BIoEx) J 1(1):18–21

Mahtab Z, Azeem A, Ali SM, Paul SK, Fathollahi-Fard AM (2021) Multi-objective robust-stochastic optimisation of relief goods distribution under uncertainty: a real-life case study. Int J Syst Sci Oper Logist 9(2):241–262

MirHassani SA, Lucas C, Mitra G, Messina E, Poojari CA (2000) Computational solution of capacity planning models under uncertainty. Parallel Comput 26:511–538

Nezhadroshan AM, Fathollahi-Fard AM, Hajiaghaei-Keshteli M (2020) A scenario-based possibilistic-stochastic programming approach to address resilient humanitarian logistics considering travel time and resilience levels of facilities. Int J Syst Sci: Oper Logist 8(4):321–347

Niño M, Vanessa P, Vizcaíno G, Libardo S (2010) Stochastic programming models for designing logistic networks. Prospectiva 8(2):13–19

Owen SH, Daskin MS (1998) Strategic facility location: a review. Eur J Oper Res 111:423–447

Powell W, Topaloglu H (2003) Stochastic programming in transportation and logistics. In: Handbooks in operations research and management science, vol 10. Elsevier, pp 555–635

Shapiro JF (2001) Modeling the supply chain. Massachusetts institute of technology, Duxbury Thomsons Learning

Zelenika R, Pupavac D (2008) Menadžment logističkog sustava, Ekonomski fakultet Rijeka, Rijeka

Zhuang X, Zhang Y, Han L et al (2023) Two-stage stochastic programming with robust constraints for the logistics network post-disruption response strategy optimization. Front Eng Manag 10:67–81

Impact of Digitalization on Ecological Sustainability in Warehousing—Case Study of JD.com and SF Express Warehouses in Shanghai

Xixin Pang and Dagmar Piotr Tomanek

Abstract Warehousing is one of the crucial elements in the distribution of goods and plays an indispensable role in modern logistics. Since the introduction of Industry 4.0, the level of digitalized warehousing has been steadily improved. In addition, the ecological sustainable development of warehousing has become one of the most important development trends of warehousing (Dohrmann et al., Logistics Trend Radar 7.0. DHL Group, (2024)). How can the impact of digitalization on ecological sustainability in the context of warehouses be quantified and presented in a meaningful way? The authors' assessment of two leading companies in China's logistics industry offers an innovative approach. The methodical procedure of this work includes a case study analysis of the two Chinese global players, namely JD.com and SF Express, in the megacity of Shanghai. JD.com is China's largest integrated e-retailer and its Shanghai Asia No.1 warehouse corresponds to the warehousing with a high level of digitalization. SF Express is a comprehensive service enterprise mainly engaged in international and national express delivery with a relatively traditional warehousing system. The results show that an increase in the degree of digitalization can lead to an increase in the degree of ecological sustainability. However, a higher degree of digitalization is required for a certain degree of increase in ecological sustainability.

Keywords Ecological sustainability · Digitalization · Industry 4.0 · Intelligent warehousing · Green logistics · JD.com · SF express · Shanghai

X. Pang · D. P. Tomanek (✉)
ISM International School of Management, Hamburg, Germany
e-mail: dagmar.tomanek@ism.de

X. Pang
e-mail: xixin.pang@ism.de

A. Kolinski and M. Adamczak (eds.), *Digitalisation of the Greening Supply Chain*, EcoProduction, https://doi.org/10.1007/978-3-031-88918-9_7

1 Introduction

Warehousing is one of the important links in the flow of goods and an important pillar of logistics activities. With the continuous high-speed development of urbanization, digitalization and ecological sustainability are increasingly becoming mayor trends in the development of urban logistics. Logistics by function is divided into transportation, warehousing and material handling. The development of the warehousing business occupies a central position in the entire goods cycle. Warehousing costs have a major influence on the development of the logistics sector. Modern logistics aims at integrating processes, coordinating upstream and downstream and emphasizes optimal inventory. Stationary robotics, indoor mobile robots, Radio Frequency Identification (RFID), advanced analytics, generative AI are already widely used in the industries like e-commerce to enable the automation and efficiency in warehouse management processes by the use of digital technologies.

General warehouse management mainly includes main processes like receiving of incoming goods, storage, order picking and shipping of outgoing goods. Intelligent warehousing system mainly includes automatic identification system, automatic handling system, storage system, automatic sorting system and warehouse management information system, corresponding to the general warehouse management of the main processes.

Currently, the existing intelligent warehousing industry is developing from the initial 1.0 version, that is, purely manual sorting to the use of automated warehousing facilities, the selection of intelligent equipment stage and ultimately to the realization of intelligent logistics warehousing solutions for the integration and implementation of the development. With the maturity of artificial intelligence, Internet of Things and other technologies are also widely used in warehousing, making intelligent warehousing more and more popular. Efficient and reasonable intelligent warehousing can not only help enterprises to speed up the flow of goods, but also reduce costs, so that enterprises in the fierce competition in the industry increase the competitive advantage and highlight their core competitiveness. In fact, intelligent warehousing has become an important part of logistics. On the one hand, its application not only makes the data in the process of goods inbound, storage, handling and outbound more effective and efficient by avoiding more human errors and accelerating the throughput of goods. On the other hand, the interconnectivity of departments within a company guarantees the truthfulness of the data.

With global environmental problems becoming more and more serious, in addition to the level of digitalization, the ecological sustainability of warehousing has likewise become an element that cannot be ignored. It is designed and applied for some green aspects of warehouse management, minimizing energy consumption and environmental pollution while ensuring warehouse operations. In the future, green warehouse management will become more important. As the green warehouse is mainly to reduce energy consumption, improve resource utilization and reduce environmental pollution, so this type of warehouse can not only achieve the purpose of environmental protection, but also can shrink the cost of enterprise expenditure, reduce the operating costs of enterprises and enhance the ecological sustainability of the warehouse.

Sustainability is one of the main objectives of developing green logistics. Under the double promotion of economic globalization and e-commerce, the ecological sustainability of logistics is rapidly transforming from traditional logistics to modern logistics and becoming an inevitable trend in the development of the current logistics industry. Under the support of digitalization technology, with information technology as the core, strengthening resource integration and optimization of the whole process of logistics is an essential feature of sustainable development of logistics. In the international field, information technology and digitalization of the two key technologies of systematic integration of the application of logistics have also played an obvious role in promoting the sustainable development of logistics, greatly reducing the cost of logistics and continuously innovating under the support of the new technology platform.

In the competition of major logistics service enterprises committed to seeking technological leadership, the development of modern green logistics technology generally shows four major trends, namely, the automation of logistics technology and equipment, the information technology of logistics operation and management, the intelligence of logistics operation processes as well as the integration of a variety of technologies and hardware and software platforms. Comparatively speaking, the information technology and integration of modern green logistics are developing rapidly at present, which is more popular in the horizontal field. In the vertical field, the development direction of modern green logistics information technology is integration and intelligence.

As the largest metropolitan city in China, Shanghai is the economic, technological, trading, financial and transportation hub of China. The logistics industry in Shanghai utilizes the most advanced technological equipment in China, which provides strong support for the ecological sustainable development of Shanghai's logistics and is more representative of China's current level of application of advanced equipment in logistics. At the same time, Shanghai, as a transportation hub in China, has attracted many logistics companies to set up warehouses and distribution centers here. Therefore, it is necessary to select appropriate research objectives. There are two reasons to select JD.com and SF as case studies for this study: on the one hand, as the leading companies in China's logistics industry, JD.com and SF do not have a huge gap in terms of scale and they are both committed to ecological and sustainable development. On the other hand, JD.com is equipped with many advanced digitalization technologies in warehousing, while SF is more inclined to traditional warehousing mode. Therefore, comparing these two logistics companies is more helpful in analyzing the impact of intelligence and digitalization technologies on ecological sustainability.

The key contributions of this work are directed towards the warehousing aspect of logistics, especially towards the level of application of digitalization technologies and the performance of ecological sustainability. Through the study of related theories, we can gain a deeper understanding of warehousing, including:

- The measurement of digitalization and ecological sustainability in warehousing,
- The impact of digitalization on ecological sustainability and
- Problems faced by the relevant cases in current warehousing and development recommendations.

At the same time, warehousing is closely linked to all parts of the logistics system. Not only in warehousing, the whole logistics system needs to be paid attention to the future development of green logistics in China's logistics industry and propose strategic measures for the implementation of green logistics.

2 The Importance of Intelligent Warehousing

In the development of economic globalization and technological support, modern "Warehousing" is no longer the traditional meaning of "Warehouse". With the rapid development of the logistics industry and production needs, it has gradually evolved into "Intelligent Warehousing" with a sense of science and technology. More and more companies are beginning to use the construction of intelligent warehousing, highly utilizing digitalization, mechanization and information technology, which gradually replaces manual labor as a way to improve the level of warehouse management and reduce costs and increase efficiency.

In a comprehensive analysis published in 2017, Choy et al. concluded that many industry practitioners find it difficult to adopt intelligent warehouse management and related methods due to complexity and high associated costs and that staff with weak artificial intelligence backgrounds still rely on experience when assigning storage locations for various products, in their 2017 article on RFID-based storage allocation systems for picking order efficiency (Choy et al. 2017).

McCrea believes that labor in warehouse management continues to maintain its place as the most expensive aspect of operational logistics. Currently, labor expenses account for about 20% of a company's costs. In 2016, workers were paid an average of $11.82 per hour for their labor. In response, many companies have shifted from manual operations to automated solutions. At the same time, some countries have implemented labor management systems (LMS), a mix of software and engineering standards that collect data on worker activity, which is considered the most effective way to help define the execution of specific tasks (McCrea et al. 2017).

By analyzing Cisco's creation, Pandian said the warehouse of the future will focus on automation and integration. There will also be the use of sensors, data collection technology, analytics and intelligently connected machines that can deliver more information from more nodes in the supply chain. Put them all together and companies get a supply chain that can automatically take action up and down the supply chain, including reacting in the warehouse when things happen (Pandian 2019).

As countries around the world began to seek intelligent warehouse management solutions, the management of China's warehouses are also changing. Despite the large number of warehouses in China, the level of management and technical level is uneven. Most of them are still limited to the function of storing goods. As a result, most Chinese warehouses generally have only a few information technologies, smart machines and advanced equipment in use. The use of advanced warehouse management models is also rare.

Some Chinese scholars have studied the important role of the development of intelligent warehousing in 2020 and through the analysis of the path of realization of the integrated system of warehousing management and control based on the Internet of Things, it has been concluded that the operational equipment in Chinese warehouses at present has a low degree of digitalization and poorly interactive.

Due to the massive manufacturing industries in China, there is no strict sense of equipment production standards, leading to the diversification of equipment, applied to the warehouse. This also makes even many of the same use of warehouse configuration equipment models, but the performance difference is huge. In the actual use of warehousing, despite the use of some of the necessary operating equipment, the different technologies of the various manufacturers have led to the emergence of poor equipment linkage, operational incoherence and other issues. In addition, these scholars believe that the application of intelligent warehouse management can effectively improve the warehouse operation ability and avoid the problems of the traditional warehouse, but only rely on the investment of mechanized equipment or the use of a set of information technology software system, can not completely avoid the problems in the actual operation and to ensure that the industry continues to move forward (Hao et al. 2020).

With the continuous application of computer technology and the development of mechanical equipment updates, it makes traditional warehouse management more modern and intelligent. In the day-to-day work, these intelligent operations can not only improve efficiency and reduce inventory costs, but also promote the industry's continuous progress.

Tong Jing and some scholars have analyzed the intelligent warehouse management system and combined the big data technology, Internet technology, information collection and the functions required in the warehouse operation in an effective and unified way and believe that in the future, with the organic combination of information collection technology, automated warehousing technology and radio frequency technology, the intelligent warehouse management system will be better able to meet the needs of the industry's development (Tong et al. 2023). Zhang has researched the development of China's warehousing and Japan's warehousing and by analyzing and comparing the overall planning and supervision of the warehousing industry as well as the innovation of management technology from the warehousing enterprises themselves, he believes that, comprehensively, with the support of national policies, China's warehousing industry has already realized the limitations of its own development and steadily begun to consciously make changes. Especially between 2013 and

2018, the level of the industry has significantly improved, the level of intelligence has gradually increased and investment in the warehousing industry once became popular (Zhang and Zheng 2021).

3 Green Logistics in China

At present, some studies on green logistics are mainly conducted from the perspectives of green transportation, sustainable development, green costs, constraints and the role of various factors in green logistics and improvement strategies are proposed.

Rodrigue and other scholars in 2017 proposed the establishment of green ecosystems and logistics and transportation for the organic combination of scientific and technological progress for the effectiveness of modern logistics, to develop a sustainable logistics system (Rodrigue et al. 2017). They explored effective measures to reduce carbon dioxide emissions in logistics and transportation from the perspectives of transportation fuels and public transportation infrastructure and put forward suggestions for improvement.

Herold et al. used a simulation method to evaluate green logistics practices in the automotive industry, integrating factors such as carbon dioxide emissions, waste disposal and energy utilization and studying the distribution of logistics (Herold and Lee 2017). Browne and McKinnon and other scholars explored the recycling and environmental logistics channel of sustainable society in the article of green logistics and considered that the green recycling environmental logistics channel of reuse and recycling plays an important role in the governance of the logistics industry (McKinnon et al. 2015). Degirmenci and other scholars analyzed the green transportation aspect in 2017 and used the SWOT model to study the development prospects of the green automobile market, emphasizing the importance of green automobile subsidies for modern logistics (Degirmenci and Breitner 2017).

Chinese scholars in the field of green logistics have mostly focused on sustainable development, constraints, green cost control and other aspects and carried out corresponding research work in the context of the actual environment of the domestic logistics industry.

Chu et al. based on the current situation of logistics development in the northern region of China, adopted the greenness regional evaluation model for green logistics evaluation, which is considered to be a relatively perfect modeling system that can be extended to the green logistics evaluation of most enterprises (Chu et al. 2019). In fact, compared with developed countries, the green logistics programs of most logistics enterprises in China are still in the trial stage and have not formed a corresponding scale effect. Some scholars focus on comparing the correlation between environmental protection and logistics transportation when studying green logistics and further elaborate on the possible pollution impacts of logistics transportation on various regions, including waste, reverse logistics and packaging materials. Based on the above studies, researchers further propose that green logistics programs in China should be implemented in a more efficient way. Based on the above research, the researcher further proposes that the participation in the development of green logistics from the perspective of enterprise managers should be based on the conditions of negative impacts of logistics transportation in the external environment so as to ensure that the relevant logistics resources are fully utilized and to achieve the goal of purifying the environment while reducing the burden on the enterprise (He et al. 2017). Liu et al. discuss the green logistics development program adapted to China's development environment, which involves packaging materials, product transportation, waste recycling and other aspects. Under the condition of the green logistics plan as a guideline, further decomposition of each transportation link can truly build a perfect green logistics service system (Liu et al. 2018).

Green logistics involves the change of transportation means, the training of core talents and the large input of green materials, which will bring cost pressure to logistics enterprises to a certain extent. Based on this feature, Chinese scholars have conducted researches on the obstacles to green logistics and Bai et al. analyzed the spatial dependence of the logistics industry in realizing green development by using carbon dioxide emissions as a green environmental indicator and concluded that the spatial dependence of green logistics development is the main limitation of green logistics (Bai et al. 2015). Some scholars have also analyzed the problems of green logistics in China from the theory and influencing factors of green logistics and considered that green logistics should be improved from various perspectives, such as government management, enterprise strategy, ideological understanding and waste recycling (Wang 2016). Some other scholars take the constraints of China's green logistics as the research direction, discussed the content of its core areas and believed that it is necessary to improve the construction of green logistics from the aspects of policy, infrastructure, environment and talent cultivation. Khan et al. analyzed the obstacles related to the process of realizing the greening of the express delivery industry in 2016 and believed that the poor acceptance of the upstream and downstream, the quality of service and the cost of transportation. Are the main constraints. The promotion of green logistics mode will lead to a large increase in transportation costs in a short term and may have a negative impact on service quality (Khan et al. 2016).

4 Measuring the Degree of Digitalization and the Degree of Ecological Sustainability

In order to measure the impact of digitalization on ecological sustainability, it is first necessary to integrate all the factors related to digitalization and ecological sustainability separately into a unified quantitative key performance indicator.

As the vast majority of digitalized technology and equipment is used for warehousing, the number of parcels handled may to some extent reflect the level of digitalization in the warehouse. Following this logic, the degree of digitalization in this study measures the number of packages handled per employee per day, which is calculated as:

$$\begin{aligned}&\text{Key Performance Indicator "Degree of Digitalization"}\\&= \text{Number of packages handled per employee per day}\\&= \frac{\text{Number of packages handled per day}}{\text{Number of employees}}\end{aligned} \tag{1}$$

In the estimation of the degree of ecological sustainability, the volume of total carbon emission reductions is considered as the measurement of the degree of ecological sustainability:

$$\begin{aligned}&\text{Key Performance Indicator "Degree of Ecological Sustainability"}\\&= \text{Total Carbon Emission Reductions per year}\\&= \sum \text{Carbon Emission Reductions [in tonne]}\end{aligned} \tag{2}$$

5 Case Study of JD.Com and SF Express Warehouses in Shanghai

Case study method is the main research method in this paper. With the launch of China's 14th Five-Year Modern Logistics Development Plan for the period from 2021 to 2025 and other policies, the ecological sustainable development of warehousing has also become one of the important development trends for logistics companies in China. Shanghai is the largest metropolitan city in China. At the same time, Shanghai is the economic, technological, commercial, financial and logistic center of China. For this reason, the city of Shanghai is an excellent example for examining the impact of digitalization on ecology in warehousing.

JD.com and SF in Shanghai were selected for the case study. For detailed justification, JD.com and SF are the leading companies in China's logistics industry. There are only marginal differences in size between the two companies, which facilitates comparability. Both companies are also committed to ecological and sustainable development. On the other hand, there are significant differences in the degree of digitalization. JD.com is equipped with many advanced digitalization technologies in warehousing. SF, on the other hand, tends more towards traditional warehousing with low digitalization. For these reasons, a comparison of the two logistics companies is useful for analyzing the impact of intelligence and digitalization technologies on environmental sustainability.

For the selected the research objects, JD.com and SF, the data of the last few years are compared and analyzed to explore the current status of their warehousing in Shanghai. During the research, relevant data from the financial annual reports, ESG annual reports, official website data, company announcements and other information disclosed by the relevant case companies. Since some of the data in the above sources are national data rather than the data in Shanghai area, they are not in line with the research object of this study. Therefore, in order to solve this issue and verify the authenticity and validity of the relevant data, telephone interviews with two companies were conducted. The basis for the data valuation was obtained from the inquiries in the telephone interviews—Jingdong and SF's warehousing data in Shanghai account for about 1/7 and 1/10 of the national warehousing data, respectively.

5.1 JD.com's and Its Status of Warehouses in Shanghai

JD.com, a Chinese self-operated e-commerce company, founder Liu Qiangdong serves as the chairman of the board of directors and chief executive officer of JD.com Group which has JD Online Mall, JD.com Finance, Paipai.com, JD.com Intelligence and Overseas Division. In 2013, it officially obtained the virtual operation license. In May 2014, JD.com was officially listed on the NASDAQ Stock Exchange in the U.S. In July 2015, JD.com was selected as a member of the Nasdaq 100 Index and the Nasdaq 100 Average Weighted Index.

JD.com is China's largest integrated e-retailer and one of the most popular and influential e-commerce websites in China, selling millions of high-quality products from tens of thousands of brands in 12 categories, such as home appliances, digital communications, computers, home furnishings, clothing and apparel, maternal and child care, books, food and online travel. In the first quarter of 2012, JD Online Mall ranked first among China's self-operated B2C websites with a market share of 50.1%. In the first quarter of 2012, JD Online Mall ranked first among China's self-operated B2C websites with a market share of 50.1%. Compared with similar e-commerce websites, JD.com Mall is characterized by providing authentic licensed products, machine-printed invoices and after-sales service, as well as "Price Protection", "Extended Warranty Service" and other quality services. With more competitive prices and gradually improved logistics and distribution system and other advantages, JD.com won the market share for many years and ranked first in the industry.

The rapid growth and rise of JD.com have attracted the favor of venture capitalists from all over the world and more than 1.5 billion USD of hot money has been invested in JD's development, which has undoubtedly injected sufficient energy into the company's development. Under the guarantee of quality capital, JD.com has continuously strengthened the construction of logistics and warehousing centers. At present, JD.com has set up six logistics centers in North China, East China, South China, Southwest China, Central China and Northeast China, which has realized the further optimization of logistics services. As of March 2023, JD.com operated more than 1,500 warehouses and the total area of JD.com warehouses, including the area managed by the cloud warehouse ecological platform, exceeded 31 million square meters. Meanwhile, as of the end of 2022, JD.com also owned nearly 90 bonded warehouses, direct mail warehouses and overseas warehouses around the world, with a total area under management of nearly 900,000 m^2. In 2022, JD.com completed the acquisition of Deppon Logistics, quickly acquiring an express transportation network covering China.

JD.com's "Asia No.1" Shanghai logistics center located in Jiading, Shanghai, as one of the modern logistics centers with the largest building scale and the highest degree of automation in the B2C industry within the Asian scope, it perfectly dispatches automated equipment such as conveyor lines, sorters, elevators, which greatly supports and promotes the logistics operation of the large platform of JD.com. The park of JD.com's Asia No.1 in Shanghai is with a floor area of about 300,000 m^2, equivalent to the size of more than 40 standard soccer fields. Through the use of a large number of three-dimensional stacking storage technology and automated sorting equipment, it has been able to handle an average daily order volume of more than 500,000 pieces. In terms of warehouse capacity, it can support a maximum of about 4.3 million pieces of merchandise storage needs.

The JD.com Asia No.1 Warehouse is divided into four areas—a three-dimensional storage area, a multi-level attic picking area, a production work area and a delivery sorting area. The "Three-Dimensional Storage Area" is 24 m high and utilizes an automatic access system (AS/RS system) to achieve automated high-density storage and high-speed picking capabilities; the "Multi-Level Attic picking area" deploys a variety of modern equipment to achieve automatic replenishment, The "Multi-Level Attic Picking Area" adopts various modern equipment, realizing automatic replenishment, fast picking, multiple checking manners and multi-level attic automatic conveying capacity, which realizes high-density storage of huge stock keeping units (SKU) and fast and accurate picking and conveying capacity of JD.com. "Production Operation Area" adopts JD.com's self-developed task assignment system and automated conveying equipment to realize automation and rationalization of task assignment for each production workstation, guaranteeing the full-load operation of each production position, avoiding the uneven allocation of tasks and greatly improving the labor efficiency. The "Delivery and Sorting Area" adopts an automated conveyor system and a sorting system that represents the highest level in the world, with a sorting accuracy rate close to 100%, which effectively solves the problems of poor efficiency and low accuracy of manual sorting.

In addition, compared with traditional warehousing, Asia No.1 has optimized the operation mode of warehouse management, simplified the work of operators and effectively reduced the possibility of errors in warehousing. Its optimized warehouse management mode of operation includes:

Mixed-zone operation strategy. Shanghai Asia No.1 covers an extensive area and it is unrealistic to operate such a large warehouse with traditional manpower. Now the warehouse is divided into 28 regions, each region arranges a staff in charge of the overall picking and packaging review. It turns the traditional division of function points into regional task management, i.e., mixed job operation mode and these operations can be done only at the workbench.

Multi-task management strategy. Based on the flow of goods and pallets, JD.com has established a multi-mode and complete order task management mechanism. Scanning goods and pallets can automate the flow of tasks instead of relying on manual task assignments and establish empty pallets, empty crates and other goods management mechanisms.

Automatic task assignment strategy. In Asia No.1, order tasks will be automatically assigned to employees, instead of people looking for tasks. Operators only need to do what actions they should do, do not need to care what the next step of the process is and the system will have a complete set of order management mechanisms so that employees can automatically get each work task.

Efficient cargo sorting strategy. In Asia No.1, goods of different orders are not merged in the warehouse, to prevent unnecessary backlogs caused by merging of orders in the warehouse, thereby minimizing the amount of goods to be picked. Sorted goods flow with the conveyor transfer to the review station and the automated equipment scans the code to confirm the cargo. The system matches orders based on a series of prioritized value strategies for commodities and in this way, the waiting time for order review can be further reduced to improve the efficiency of warehouse operation.

Based on the algorithm, JD.com's Shanghai Asia No.1 has implemented a more reasonable warehouse layout and efficiently utilized digitalization equipment, greatly reducing manual operations and improving the rationality and accuracy of the entire warehouse operation.

5.2 SF Express and Its Status of Warehouses in Shanghai

Established in 1993 and headquartered in Shenzhen, SF Express is a comprehensive service enterprise mainly engaged in international and national express delivery and related fields. Since its establishment in 1993, SF Express has always been focused on quality and service enhancement to meet the market demand continuously and has established a considerably huge business organization in the Greater China region (including Hong Kong, Macao and Taiwan) for market development, information collection, express delivery and receipt, logistics and delivery. Furthermore, SF Express has built up a nation-wide and even a global network to serve its customers.

For a long time, SF has invested a lot of money to strengthen the company's basic construction, actively introduce and develop domain name registration query equipment with high-tech content and information technology, continuously improve the level of automation of warehousing operations and realize the information query, monitoring, tracking and scheduling of resources for the whole process of logistics flow, which promotes the continuous optimization of the SF Express network and ensures the steady improvement of the level of service, laying the foundation for customer service satisfaction in the industry. It has promoted the continuous optimization of the SF Express network, ensured the steady improvement of service level and established the leading position of customer service satisfaction in the industry.

As an important link in the downstream of e-commerce, the express delivery industry in just a few years achieved rapid expansion from an obscure fringe industry to stand in the spotlight, the national express delivery market into the express delivery business volume and business income are growing at a high rate. Although the express industry has entered a new development channel, the industry's own limitations restrict the customer service personnel career development and development of the express industry. In order to encourage employees, SF has adopted a set of "Their Own Boss" pay scales. SF Express's relevant person in charge of the company introduced, SF using piecework, each courier is their own boss and the employee's compensation all depends on diligence and customer recognition.

In 1996, with the steady growth in the number of customers in the industry and the booming of the country's economic megatrend, SF Express further expanded its network to major cities across the country beyond the provinces. Up to now, SF's express service network has covered more than 20 provinces and municipalities in China, more than 100 prefecture-level cities, also including the region. It has become the leader of the private brand in China's courier industry. SF Express, as a Hong Kong-funded courier company that mainly operates international national courier business, provides fast, safe, economic, accurate and high-quality courier services for the majority of customers. SF has always been to promote the economy, the achievement of the customer and the development of the national courier industry as its own mission, actively explore the needs of the customer and constantly launch the promotion of new courier and related business service programs to provide customers with fast, safe and accurate distribution channels. In order to provide customers with safer and more convenient courier services, SF Express network adopts a self-built and self-operated business mode. After more than twenty years of continuous development, SF Express has more than 100,000 employees, more than 30 branches and more than 18,000 self-built delivery stations and its service network has covered more than 20 provinces, municipalities directly under the central government and regions, more than 100 prefectural-level cities in China.

In recent years, SF warehousing development momentum has been positive, due to changes in supply and demand in the logistics market, SF warehousing is developing in the direction of warehousing socialization, warehousing industrialization, warehousing standardization, warehousing modernization, SF has become a logistics enterprise with advanced management concepts, began to focus on the management of the overall supply chain and actively develop long-term relationships with customers. SF Express participates in customers' supply chain management by providing various value-added services and reducing customers' costs, which can also improve its own competitiveness. At the end of the twentieth century, as China's economic development reached a high level, it also drove the development of SF warehousing, which put forward higher requirements for SF warehousing development.

At the same time, along with the development of e-commerce in recent years and the rise of Online to Offline (O2O) business model, the whole society for e-commerce logistics requirements are increasing. Along with consumption upgrading and the awakening of consumer awareness, e-commerce logistics is further required to have better service and timeliness and a variety of logistics modes and technological innovations have emerged. SF, as a leading logistics company has built its own logistics system by adopting the warehousing-distribution logistics mode to satisfy the offline logistics service of the online platform. Warehousing-delivery logistics is one of the main ways of e-commerce logistics at present.

Warehousing-delivery logistics mode takes the former warehousing as the means, utilizes big data to predict the sales volume of goods in each region and stores a certain amount of goods in advance in the warehouse closest to the consumers. After the order is issued, the goods are then picked up and delivered from the warehouse closer to the consumer, shortening a large amount of transportation and transit time and significantly improving distribution timeliness. SF Express has always been recognized by the majority of consumers for its high timeliness characteristics.

The core driving force of SF Express is technology and data. The company has taken the enrichment of an infrastructure support system as its business goal, specialized warehouse and distribution services, information technology platform construction and other aspects of leapfrog development, efficiently responding to the diversified demand for supply chain services from customers in various fields and creating intelligent logistics parks to help its partners reduce costs and increase efficiency.

Although SF Express has always been at the forefront of the logistics industry, self-owned traditional warehousing centers still occupy a dominant position, there are still problems, such as non-standardized warehousing processes, chaotic management of warehousing product sorting, low operational efficiency and poor personnel management. SF's warehousing center in Jiading District, Shanghai, is a traditional warehousing center, which currently has more than 1,100 employees and covers an area of more than 60,000 m^2. The center has more than 1,100 employees and covers an area of more than 60,000 m^2. The warehouse center is mainly for SF online platform sales of products for storage and storage and sorting and packaging delivery operations. SF Shanghai warehouse center of the random arrangement of the various functional areas, the storage area covers an area of the largest, but the placement of its goods are randomly arranged. A lot of time are wasted and many problems in the warehouse are caused by this.

5.3 *Measuring the Degree of Digitalization*

In order to measure the degree of digitalization of the two companies, it is necessary to collect and organize the data from corporate materials, which mainly include ESG Annual Reports, Financial Annual Reports, corporate announcements as well as data and information published on the official websites. During the selection of relevant data, it is inevitable that many data stated are nation-wide and not all the data regarding the Shanghai warehouses of the two companies are disclosed in these materials. In order to obtain the relevant data regarding Shanghai warehouses, telephone interviews were conducted and solutions were proposed during the telephone interview with customer services from the two companies—the relevant data could be estimated based on the ratio of resources in Shanghai to that in China.

According to the suggestions given by the interviewees, the resources of JD.com's Asia No.1 Shanghai warehouse account for about 1/7 of JD.com's national warehouse resources and the resources of SF's Shanghai warehouse account for about 1/10 of SF's national warehouse resources. Based on the estimation of these two ratios, the data of the relevant digitalization technologies applied by JD.com and SF in their Shanghai warehouses are shown in Table 1.

From Table 1, it can be found that there is a considerable gap between the digitalization level of JD.com and SF warehouses. First of all, in JD.com's Asia No.1 in Shanghai, AGVs are widely used and the park can operate about 800 AGVs at the same time. There are three types of AGVs in the park, which are classified into large, medium and small according to their models. Small and medium-sized AGVs operate inside the picking track to transport the goods, while large AGVs directly transport the cargoes to different delivery ports after the cargoes fall into the collection ports. Inside the sorting yard, JD.com deploys different visual recognition technologies and infrared ranging technologies to realize their active interaction with the environment. These AGVs take 10 min to recharge each time and deliver goods according

Table 1 Comparison of warehouse digitalization technologies between JD.com and SF; Sources: own study based on JD.com (2022b, a) Environmental Social and Governance Report, JD.com (2022b, a) Annual Report; SF (2022a, b) ESG Report, SF (2022a, b) Financial Report, Corporate Official Website of JD.com & SF and Telephone Interviews

JD.com Asia No.1			SF		
Digitalization technologies	Number	Unit	Digitalization technologies	Number	Unit
AGV	≈800	/	/	/	/
JD shuttle system (in three-dimensional area)	8	Sets	/	/	/
6-axis Robotic Arm	73	Sets	/	/	/
New energy forklift trucks	>100	/	/	/	/
New energy logistics vehicles (for park transportation)	>1,100	/	New energy logistics vehicles (for park transportation)	≈2,100	/
Photovoltaic power generation equipment	>100,000	m^2	Photovoltaic power generation equipment	>9,000	m^2
Automated green packaging recycling	≈7,857,000	Pieces	Automated green packaging recycling	≈2,774,000	Pieces

to different tracks. In case of urgent cargoes, other AGVs will automatically make way for the urgent cargoes to be delivered first.

In addition to AGVs, Asia No.1 has deployed 8 sets of JD Shuttle, a highly intelligent "Goods-to-Person" system, which mainly performs the functions of warehousing, storage and picking, with a total of about 43,000 crate storage spaces and 6 picking workstations. The picking workstations are ergonomically designed, with the crates placed at an inclination of 30° and the error-free picking concept is used to manage the operation process, with an operational efficiency of up to 1,000 grips per person per hour. The shuttle vehicle carries out storage and picking operations through the aisle and its storage density is more than three times that of conventional shelves. Compared with conventional manual picking, the efficiency has been increased by 6–8 times. JD Shuttle intelligent system consists of shuttle vehicles responsible for handling in parallel directions and elevators responsible for handling in up and down directions.

In the warehousing segment, JD.com's Asia No.1 campus deploys 73 six-axis robots for warehousing operations. The six-axis robots are about 2 m tall, with the largest arm span of about 2.7 m and weighing about 1.7 tons and are characterized as being suitable for picking up large items. The six-axis robot can handle goods weighing up to 165 kg and its handling efficiency is high, as it takes 8 to 12 s to move a word. Compared to manual work, the handling efficiency can be increased by 30%.

Table 2 shows the data of JD.com's Asia No.1 Shanghai warehouse and SF's Shanghai warehouse in terms of warehousing and employees. From the table, it can be observed that JD.com's Asia No.1 in Shanghai occupies about 300,000 m^2, which is almost equivalent to the size of five SF's Shanghai warehouses. In terms of cargo stocking capacity, due to the difference in logistics models between JD.com and SF, JD.com needs more stocking capacity in its warehouses, while SF's warehouse is more likely to play the role of a transit station. As a result, JD.com's Asia No.1 warehouse in Shanghai also has a much larger maximum storage capacity than SF's, which is roughly 4,300,000 pieces, while SF's warehouse in Shanghai can store up to about 65,000 pieces of goods at the same time.

With the assistance of AGVs, Asia No.1 is also more efficient at sorting goods than SF's Shanghai warehouse. According to official statistics from JD.com and SF in 2022, JD.com's Asia No.1 in Shanghai has a sorting efficiency of about 35,000 pieces per hour, compared to SF's 11,000 pieces per hour. Currently, Asia No.1 can handle more than 500,000 pieces of packages per day, a volume that requires more than 600 people to work simultaneously in the sorting scene alone. However, with various digitalization equipment working together, the number of employees in Asia No.1's campus is only about 480; SF's Shanghai warehouse handles about 235,000 parcels per day and the warehouse campus contains about 1,100 employees.

Combining the data in Table 2 with the calculation of the key performance indicator "degree of digitalization" (1), the degree of dogitalization of JD.com's Shanghai Asia No.1 and SF's Shanghai warehouse is visualized in Fig. 1. JD.com's Asia No.1 in Shanghai is not only significantly more efficient than SF's Shanghai warehouse,

Table 2 Comparison of warehouse effectiveness between JD.com and SF; Sources: own study based on JD.com (2022b, a) Environmental Social and Governance Report, JD.com (2022b, a) Annual Report; SF (2022a, b) ESG Report, SF (2022a, b) Financial Report, Corporate Official Website of JD.com & SF and Telephone Interviews

JD.com Asia No.1			SF		
Warehousing effectiveness	Number	Unit	Warehousing effectiveness	Number	Unit
Floor space of the warehouse park	≈300,000	m^2	Floor space of the warehouse park	≈60,000	m^2
Maximum cargo stocking capacity	≈4,300,000	Pieces	Maximum cargo stocking capacity	≈65,000	Pieces
Warehouse sorting efficiency	≈35,000	Pieces/hour	Warehouse sorting efficiency	≈11,000	Pieces/hour
Number of staff in park	≈480	/	Number of staff in park	≈1,100	/
Average daily quantity of cargo handled in warehouse	>500,000	Pieces/day	Average daily quantity of cargo handled in warehouse	>235,000	Pieces/day

but also has less than half of SF's employees. Consequently, the degree of digitalization of JD.com's Shanghai Asia No.1 is almost 5 times higher than SF's warehouse in Shanghai (see Fig. 1).

5.4 Measuring the Degree of Ecological Sustainability

Since SF has invested much more in logistics and transportation equipment than JD.com, it is clear from Table 3 that SF has about twice as many new-energy vehicles in its Shanghai warehouse park as JD.com. SF has about 2,100 new-energy vehicles in its warehouse park for intra-warehouse transportation, while JD.com has only about 1,100 vehicles. Comparing the floor space of JD.com's Asia No.1 warehouse in Shanghai with that of SF's warehouse in Shanghai, JD.com's Asia No.1 is about 5 times the size of SF's warehouse and its deployment of new energy vehicles is only about 1/2 that of SF's. It is evident that SF's deployment of new energy vehicles is far denser than that of JD.com.

Both JD.com and SF have installed photovoltaic power generation facilities in their warehouse parks. JD.com has installed 9 MW of photovoltaic power generation facilities in Shanghai Asia No.1, with more than 100,000 m^2 of photovoltaic roofs occupying more than one-third of the park's total area. These photovoltaic power generation facilities have already been connected to the grid, providing Asia No.1 with a continuous supply of green energy. The green power can be used to light the office areas of the park during the day and to charge the new energy logistics vehicles at night, while SF has also equipped more than 9,000 m^2 of photovoltaic

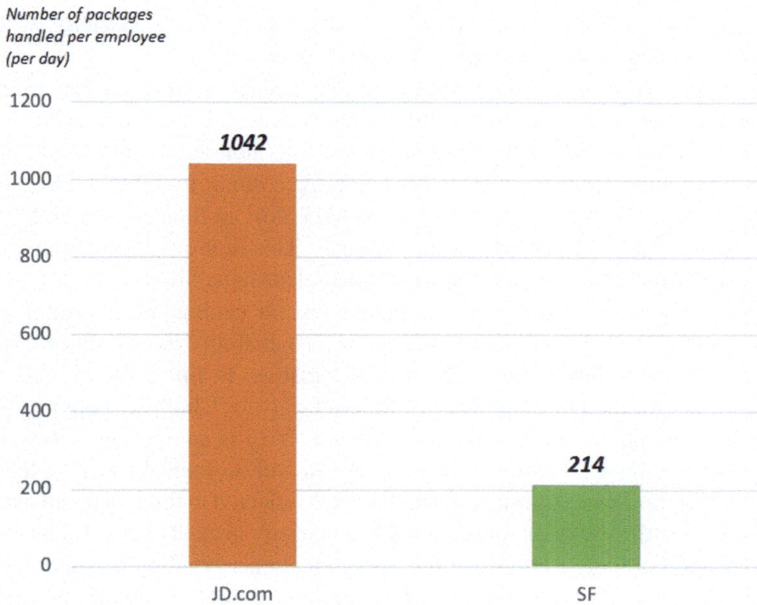

Fig. 1 Degree of digitalization. *Source* Own study

Table 3 Comparison of carbon emission reduction from JD.com and SF's warehousing operations in 2022; Sources: own study based on JD.com (2022b, a) Environmental Social and Governance Report, SF (2022a, b) ESG Report, Corporate Official Website of JD.ccm & SF and Telephone Interviews

JD.com Asia No.1			SF		
Ecological sustainability (Carbon emission reductions)	Volume	Unit	Ecological sustainability (Carbon emission reductions)	Volume	Unit
Photovoltaic system power generation = Coal savings = Carbon emission reductions	≈10,200 ≈3120 ≈6912	MWh t t	Photovoltaic system power generation = Carbon emission reductions	≈930 ≈629	MWh t
New energy vehicles*	≈23,571	t	New energy vehicles**	≈31,000	t
Green packaging Recycling*	≈88,773	t	Green packaging recycling**	≈27,000	t
Lights-out operation in warehouse	≈7260	t	/	/	/

power generation equipment as one of the park's sources of electricity, which is mainly used to charge its new energy logistics vehicles.

Last but not least, both companies have recycled their green packaging in the warehouse and there is a significant difference between JD.com's Asia No. 1 with automated packaging recycling equipment and SF's non-automated recycling and JD.com's recycling strategy is more proactive. According to Table 1, JD.com has recycled about 7,857,000 green packages at Asia No.1 in Shanghai in 2022, while SF has recycled about 2,774,000 green packages. This figure will also open the gap between the two companies in terms of ecological sustainability.

As the impact of economic development on the ecological environment has become more and more significant, ecological sustainability has become an important concept in the development of modern logistics. In line with China's "14th Five-Year Plan for the Development of Modern Logistics", both JD.com and SF are committed to corporate social responsibility. Asia No.1 is an intelligent low-carbon logistics park built by JD.com to realize the green and ecological sustainable development of warehousing. Its low-carbon efforts are reflected in three main areas: First, it will maximize the use of on-site renewable electricity instead of traditional energy; second, it will continue to improve the operational energy efficiency and electrification of the park; and third, it will offset the unavoidable residual emissions of the park through carbon reduction efforts. Similarly, SF strives to apply low-carbon technologies to the Warehouse Park in Shanghai. In order to measure the degree of their ecological sustainability, this study has collected data for the two companies' Shanghai warehouses in terms of electricity, new energy equipment and green packaging recycling (see Table 3).

Combining the data in Table 3 with the calculation of the key performance indicator "degree of ecological sustainability" (2), the degree of ecological sustainability of JD.com's Shanghai Asia No.1 and SF's Shanghai warehouse is visualized in Fig. 2. In order to ensure comparability, the size of the warehouse was taken into account when calculating the degree of ecological sustainability. In comparison to JD.com, SF has several warehouses in Shanghai. More specifically, when calculating SF's total carbon emissions, the number of digital and smart equipment in the different warehouses was combined.

From the estimation results in Fig. 2, it can be observed that the degree of ecological sustainability of JD.com Shanghai Asia No.1 is more than 2 times higher than that of SF's warehouse in Shanghai.

5.5 *Impact of Digitalization on Ecological Sustainability*

From the results of the previous data analysis, it can be concluded that the higher the degree of digitalization, the higher the degree of environmental sustainability. Consequently, the assumption that a higher degree of digitalization can lead to a higher gain in environmental sustainability is confirmed. However, the increase in the degree of digitalization does not lead proportionally to an increase in the degree of

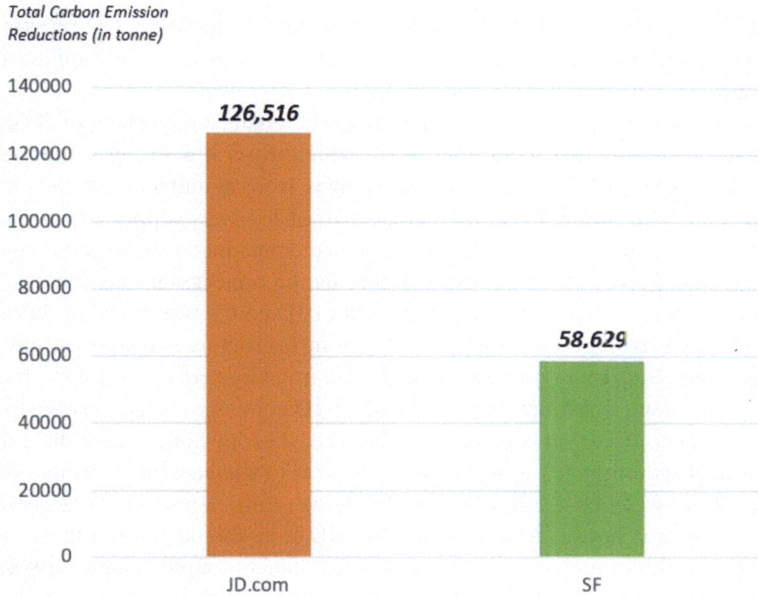

Fig. 2 Degree of ecological sustainability. *Source* Own study

environmental sustainability. It can be stated that increasing the degree of ecological sustainability requires more optimization of digitalization. In the example of the analyzed companies JD.com and SF, the ratio of the degree of digitalization is rounded 5 to 1, while the degree of environmental sustainability is slightly more than 2 to 1.

From the perspective of ecological sustainability itself, green packaging recycling can maximize its benefits and should be used as a significant reference by logistics companies for the development of ecological sustainability. Specifically, the following recommendations can be made for warehousing development at JD.com:

1. Strengthen the control of cargo inbound inspection. JD.com logistics warehouse management must be effective management of JD.com's internal operations, the inventory of goods in the warehouse and must ensure the integrity of the inventory and its authenticity. The cargo inbound should be cleaned and inspected for each of the goods, according to the results of the inspection of the quality inspection department will be the original supplier of goods electronically stamped to confirm the quality and safety of the product and the original seals will be sent to the area awaiting inspection. The quality check of the goods must be inspected by the professional testing equipment of the quality inspection department and the inspection process can not be slack. If the inspection process found that the quality of the product is seriously problematic, it must be notified to the relevant staff at the first time in the system to deal with, so as to avoid affecting the process of goods transportation. JD.com's warehousing and logistics managers should promptly resolve the uncertainties that arise in goods inbound and make

effective responses to these situations and should issue timely test reports after the completion of the quality inspection and submit them to the supplier in the system.

2. Optimize warehouse delivering capability. The greatest advantage of JD.com is that it has a large distribution network, which gives it a strong selling power. For this reason, JD.com can buy out products from manufacturers, thus having an advantage in price. From the perspective of logistics supply chain management, enhancing the current warehouse management methods, strengthening the management level of warehousing and reducing unnecessary cost waste is the inevitable stage of the future development of JD.com's warehousing. When the warehouse is under construction, the use of infrastructure and some information technology equipment not only brings JD.com a large cost investment, but also due to the long construction time, making JD.com's advantageous position in the market competition is reduced. Therefore, JD.com can increase the delivery personnel according to the actual situation so as to alleviate the distribution pressure of the warehouse and enhance the delivery efficiency after the goods are out of the warehouse. At the same time, JD.com should re-plan the scope of functions of the warehouse department, strengthen the supply chain network, the management of goods purchasing, warehousing, packaging and delivering by providing a strong guarantee for the warehouse operations.
3. Improve the comprehensive quality of warehousing personnel. For some historical reasons, China's logistics industry in such aspects of the cultural level of the staff are not very high, restricting the development of China's logistics and warehousing technology. It is a cause that logistics and warehousing technology cannot be well developed and a variety of advanced equipment remain unused. First step in developing an optimized warehousing strategy, the responsible management department should conduct regular trainings, that employees are kept up to date, timely understanding of the latest enterprise management and understand their own responsibilities, to cultivate their loyalty to the enterprise. Secondly, conduct courses of instruction for knowledge and skills improvement, so that employees can skillfully acquire the operation and storage methods of each type of equipment in the enterprise warehouse and understand the performance of each type of equipment as well as the maintenance methods in the later stage. Furthermore, through training, the employees can also gain insight into various information about the products such as quantity, variety, quality and type. In addition, in the handling of goods should also be implemented on the staff training, so that employees can be based on the characteristics and nature of the product, to find the appropriate handling tools and means of operation and avoid damage due to improper transportation, resulting in the loss of property and the quality affected by the cargo falling from shelves. Fourth, strengthen safety education. In the warehouse management sector it is often necessary to use large forklifts. If used them incorrectly, it is easy to pose a threat to safety and also cause damage to the goods, triggering the loss of property. In practice, such accidents occur occasionally, so it is important to strengthen safety education. When using forklifts and other facilities, they must be operated in accordance

with standardized actions and penalties shall be set for employees who violate them. Fifth, the team consciousness of employees should be cultivated. In warehouse management, all links are interconnected and not unrelated. If there is a problem in one segment, it will hinder the subsequent segments. It may also cause more serious consequences, triggering consumer dissatisfaction and affecting its external image. In this regard, JD.com needs to take certain measures to improve the teamwork consciousness of its employees and can hold some group activities to strengthen the teamwork consciousness of its employees.

4. JD.com should further strengthen its information management level, so as to guarantee the accuracy of warehouse management information and improve the efficiency of warehouse management. Improve the management ability of the outbound and inbound products. At the same time to strengthen the training of personnel, making the staff in line with the management of modern information technology. The use of information management will also make the level of enterprise warehouse management greatly improved, the cost down, the unnecessary loss will also be minimized. First of all, the warehouse management department adopts information management, dividing each person's responsibility specifically with a clear division of work. Improving the level of information management, so that the efficiency of information extraction greatly improved, as well as making the flow of information in various departments to speed up. Secondly, optimizing the business process of warehousing and consolidating the business process of enterprise warehousing management, thus ensuring the consistency of the business process as well as providing a guarantee for the improvement of the business. Moreover, the establishment of the information system can store all the information related to the warehousing department in the management system, which makes the sharing of information possible. When other management departments need to use the warehousing information, they can timely retrieve it from the information system, making the work much more efficient. Last but not least, the use of information systems makes it easier to process data. In case of data errors, it can be found in time through the information system and simultaneously, it makes the management of funds and the synchronization of the management of the cargo inbound and outbound. The information system can further strengthen the management of warehousing and realize the monitoring of the entire process of warehousing, timely identifying and solving problems. As a result, the competitiveness of the enterprise in the market is further improved and the development of the enterprise is promoted.

Warehousing development recommendations for SF are:

1. Optimize the layout of warehousing and fully utilize the local advantages. Many regional express delivery companies are located in urban and relatively economically developed areas with good access to the road infrastructure and main transportation routes. By adjusting their business strategy and structure, SF Express should take full advantage of the city's active information and business flow to use the benefits of the existing transportation networks. Through effective planning and multiple forms of economic cooperation, SF Express will be able to

solve many of their problems. By cooperation, SF can also improve the interior layout of the warehouses and the use of warehouse technologies, which is in line with the company's development.

2. Strengthen the technological transformation and warehouse management. The development of SF requires modern warehouse management, where information and digitalization-guided technologies have become an effective way to modernize warehouse management. Digitalized and networked computer management and the management of automated, mechanized operations will replace traditional manual operations. Warehouse management transformation mainly consists of two aspects: First, determining the transformation and construction of warehousing facilities according to the market supply and demand, thus improving warehousing automation, digitalization, intelligent transformation and enhancing the efficiency of warehousing management and the effective use of warehousing resources. Secondly, constructing a powerful and perfect warehouse management system and improve the warehouse management network and digitalization technology transformation. SF should utilize warehouse management and logistics information technology systems and digitalization technologies in all aspects of warehouse management, such as cargo sorting, warehouse management and inventory control, which can greatly improve the efficiency of warehouse management and reduce errors. Meanwhile, SF can also realize information sharing between partners and their own warehousing information through enterprise extranet and intranet. Nowadays, bar code technology, radio frequency identification technology, data acquisition technology and scanning technology are increasingly used in stacking machines, forklifts, conveyor belts and automatic guided vehicles and other equipment, mobile robots are also used as flexible logistics tools in warehousing, flexible production and product delivery plays a significant role. JD.com is way ahead of SF in this regard. Continuously improving the standardization system of warehousing management and taking the road to internationalization should be adapted to the requirements of the open market. To make proper preparations for convergence with international practice, the key is to introduce advanced management methodology and technologies, strengthen the effective integration of warehousing resources and constantly improve and develop the digitalization and intelligent system of warehousing.
3. Accelerate the application of advanced technologies. To achieve effective inventory control and improve warehouse utilization, SF must establish a sound logistics information network platform, realize the sharing of warehousing information resources and actively promote the enterprise's warehousing management level of digitalization technology. Using modern information technology to build a public information platform, complete the effective combination of warehousing information network and public information network resources, so as to enhance the level of information technology of enterprise warehousing. Using modern information technology to build a public information and data platform, complete the effective combination of warehousing information network and public data network resources, so as to enhance the level of information

technology of enterprise warehousing. SF Express can start with the following aspects:

a Introduction of advanced equipment. Equipment management is an indispensable part of enterprise management and plays an important role in improving enterprise competitiveness. Strengthening the equipment management and improving the integrity rate and utilization rate of the equipment can increase the efficiency of warehousing, reduce the cost of warehousing management and the enterprise can get more profits and achieve better economic benefits. Improvement of equipment management level will also enhance the management level of enterprises.

b Strengthen the Department of Warehousing and information communication among different departments. In enterprise management, if the communication and coordination between departments is out of balance, it will often cause the enterprise to be halted, if the information transfer between departments is disconnected, it will seriously affect the company's credibility. The establishment of the information sharing platform can assist SF departments to directly access the storage situation of the Warehouse Department and realize the informatization of warehouse management.

c Actively promote information technology, networkization and digitalization. With the rapid development of the international logistics industry, a large number of digital technologies have been adopted to improve the efficiency and quality of service in the industry. The development of modern logistics industry needs to modernize the warehouse management support, digitalization and digital-guided technologies have become the effective way to modernize the warehousing industry.

4. Strengthen the warehousing talent training and staff training. Talent is an enterprise to establish a competitive advantage, to grasp the key development opportunities, the recruitment of professionals serving the enterprise to create greater value, so that the warehouse management operations and the management level can be improved. The development of the warehousing industry, not only need to acquire a certain degree of technical experts, but also need to be good at the use of talent. However, SF currently needs more talent in these areas. Therefore, SF needs to introduce and cultivate warehousing experts and improve the training system. Simultaneously, SF should make full use of all kinds of resources, actively introduce relevant talents from various universities and strengthen on-the-job training for the staff, thus fully mobilizing the enthusiasm and creativity of the vast number of warehousing personnel through on-the-job training. In this way, SF could help employees master the application of technologies and management techniques and constantly improve the business level and cultural level of its warehousing staff.

6 Conclusions and Limitation

Accompanied with the pressure of the overall economic downturn in China after COVID-19 and the increasingly competitive market pattern, the enterprise's cost reduction and efficiency have become particularly important. Whether it is a manufacturing enterprise or a third-party logistics service enterprise, the ability to provide customers with personalized services, whether it is a manufacturing enterprise or a third-party logistics service enterprise, it is crucial to be able to provide customers with personalized services, control the inventory to save costs and improve the integration of material flow and production. Especially how the traditional logistics industry can access the core competitiveness and how to create their own digitalized warehousing system, so as to achieve the optimal design of the warehousing system, the optimal control of inventory and the most reasonable warehousing operations, will provide new impetus for the development of logistics enterprises.

At present, most of the enterprises' digitalization level of their warehouses in Shanghai is still relatively low, many are still in the state of manual work, even with the use of some information technology software systems and other digitalization equipment, but still far from being able to meet the real-time warehousing requirements. In fact, the backward traditional warehousing has affected the production operation of some companies due to the waste of human resources and low labor efficiency. For example, in SF's Shanghai warehouse, it relies entirely on human operation, which not only low efficiency, but also greatly increases the error rate, affecting the speed of goods turnover and restricting the reduction of inventory costs. Compared with traditional warehousing, intelligent warehousing has the characteristics of real-time inventory inquiry, which can improve the inventory turnover rate, reduce the labor intensity of operators and improve the accuracy of orders. The majority of traditional logistics companies represented by SF in the case study, for example, are in urgent need of warehouse intelligent upgrading, through digitalization technologies and equipment to achieve efficient and accurate warehouse management, as well as through the high-level shelves and reasonable zoning to increase the storage capacity of the warehouse. At the same time, the reasonable division of working areas makes sorting, delivery, distribution and other operations processes closely linked, which achieves the rapid operations of warehousing. The adoption of digitalization equipment will also greatly release the manpower and the cargo inbound outbound and other tasks in peak season can be efficiently completed, thus avoiding overtime for catching up tasks. Behind all this, digitalization technologies in turn provide a reliable technical guarantee for this series of operations. In logistics enterprises, warehousing is closely related to daily production and operation and is not cut off from each other. Excellent warehouse management, not only can effectively cooperate with other segments to achieve efficient management and utilization of resources, but also in other operational scenarios can achieve higher efficiency. At present, most logistics companies can not achieve real-time query of data and other information, once the realization of digitized warehouse management, this problem could be properly resolved and the joint development of various departments will be truly assisted.

The analyzed case study explores the impact of digitalization on the degree of ecological sustainability in the warehousing process. From the results of the comparison, it can be noted that the higher degree of digitalization could lead to a higher level of ecological sustainability. As problems of China's environment have been increasingly emphasized, the Chinese government has introduced policies such as the 14th Five-Year Plan for the development of the modern logistics industry, urging logistics enterprises to implement the new development concepts of innovation, coordination, greenness, ecological sustainability and openness. Companies should synergize not only warehouses, but also the entire upstream and downstream logistics system in order to develop green logistics. In order to effectively develop green logistics and promote rapid economic development, it is necessary to save energy consumption, maintain ecological balance, reduce pollution and improve the effective utilization of resources, which need government, enterprises, logistics companies and consumers to work together. As logistics companies, they need to protect the environment, reduce energy loss and rationally utilize effective resources, which can also improve the quality of employees and business level of the enterprise, thus further enhancing the economic efficiency of the enterprise. The implementation of green logistics will help form a positive business image and enhance the enterprise's sense of social responsibility. At the same time, through the requirements of green logistics, enterprises can improve the practical and technological soundness of warehousing, control environmental pollution, save energy, utilize resources efficiently, reduce production costs, reduce enterprise risks and improve the internal environment. Therefore, logistics enterprises in the future should actively improve the investment of digitalization technologies and green ecological sustainable technologies, not only to strengthen the application of advanced technology and research and development, but also to pay attention to the training of employees, so as to maximize the advantages of the relevant technology, improve logistics efficiency and further promote the ecological sustainable development of logistics. Actively promoting green logistics is beneficial for society, enterprises and individuals and it also responds to the requirements of natural development, improves social and economic efficiency and people's living environment and ensures sustainable social and economic development.

In order to investigate the impact of digitalization on ecological sustainability in warehousing, the described case study selects two companies, JD.com and SF, with their warehouses in Shanghai as the main research objectives. JD.com and SF represent high digitalization level warehousing and traditional warehousing with relatively low digitalization levels, respectively, which can serve as a good control group. To measure the degree of digitalization versus the degree of ecological sustainability, although some relevant data were indeed collected for the analysis, there are however, still some limitations in the data sample. First, many statistics of the data collected from the two enterprises are not based in Shanghai, but are nation-wide. To resolve this issue, telephone interviews with JD.com and SF's customer service were conducted. During the telephone interview, the interviewees offered their resource ratios of warehouses in Shanghai to that in the whole state. However, the ratio is only a rough estimate and is not precise enough, thus, it may affect the accuracy of the data analysis results. Second, as part of the supply chain, warehousing does not operate independently. The upstream of warehousing also has a certain impact on warehousing, for example, SF's warehouse serves more as a distribution center and their warehouse management system reserves storage space according to the number of goods sent from upstream and so on. Since the warehouse management systems of different companies are highly customized, highly differentiated and difficult to quantify, they are not researched in depth in this study.

With the proposal of the Industry 4.0 concept, more and more advanced technologies are being used in the logistics industry. Although some technologies such as RFID and AGVs have already been mentioned in this paper, there are still many more smart technologies that can be used to improve the efficiency of warehousing and the entire logistics system. Some of them are big data, digital twin, cloud computing, Internet of Things. All these technologies can further strengthen the information flow and demand forecasting in warehousing. Furthermore, blockchains can improve the synergies between warehousing and its supply chain upstream and downstream. The continuous improvement of efficiency in logistics leads to a reduction in waste, which in turn leads to a further development of green logistics.

References

Bai C, Sarkis J, Dou Y (2015) Corporate sustainability development in China: review and analysis. Ind Manag Data Syst 115(1):5–40

Choy KL, Ho GT, Lee CKH (2017) A RFID-based storage assignment system for enhancing the efficiency of order picking. J Intell Manuf 28:111–129

Chu Z, Wang L, Lai F (2019) Customer pressure and green innovations at third party logistics providers in China: the moderation effect of organizational culture. Int J Logist Manag 30(1):57–75

Degirmenci K, Breitner MH (2017) Consumer purchase intentions for electric vehicles: Is green more important than price and range? Transp Res Part D Transp Environ 51:25–41

Dohrmann K et al (2024) Logistics trend radar 7.0. DHL Group. https://www.dhl.com/de-en/home/innovation-in-logistics/logistics-trend-radar.html

Hao J, Shi H, Shi V, Yang C (2020) Adoption of automatic warehousing systems in logistics firms: a technology–organization–environment framework. Sustainability 12(12):5185

He Z, Chen P, Liu H, Guo Z (2017) Performance measurement system and strategies for developing low-carbon logistics: a case study in China. J Clean Prod 156:395–405

Herold DM, Lee KH (2017) Carbon management in the logistics and transportation sector: an overview and new research directions. Carbon Manag 8(1):79–97

JD.com (2022) Environmental social and governance report, https://ir.jd.com/system/files-encrypted/nasdaq_kms/assets/2023/06/20/18-39-49/JD.com%2C%20Inc.%202022%20Environmental%2C%20Social%20and%20Governance%20Report.pdf. Access 15 Aug 2024

JD.com (2022) 2022 Annual report of JD.com, https://ir.jd.com/static-files/9cc8985d-be5d-476b-81c5-5a0366f09d6c. Access 15 Aug 2024

Khan SAR, Dong QL, Yu Z (2016) Research on the measuring performance of green supply chain management: In the perspective of China. Int J Eng Res China 27:26–58

Liu J, Yuan C, Hafeez M, Yuan Q (2018) The relationship between environment and logistics performance: evidence from Asian countries. J Clean Prod 204:28–51

McCrea B (2017) Next-gen WMS: 7 innovations on the horizon. Logistics Manag (Highlands Ranch, Co.) 56(3):30–33
McKinnon A, Browne M, Whiteing A, Piecyk M (Eds) (2015) Green logistics: improving the environmental sustainability of logistics. Kogan Page Publishers
Pandian DAP (2019) Artificial intelligence application in smart warehousing environment for automated logistics. J Artif Intell Capsule Netw 1(2):63–72
Rodrigue JP, Slack B, Comtois C (2017) Green logistics. In: Handbook of logistics and supply-chain management (Vol 2, pp 339–350). Emerald Group Publishing Limited
SF (2022) ESG report (in Chinese), https://staticpacific.blob.core.windows.net/press-releases-attachments/1504119/1cb53c25-314e-4193-bd77-aabdc1884251.PDF. Access 27 July 2023
SF (2022), Financial Report (in Chinese), https://www.sf-express.com/uploads/2022_db01775145.pdf. Access 15 Aug 2024
Tong Q, Ming X, Zhang X (2023) Construction of sustainable digital factory for automated warehouse based on integration of ERP and WMS. Sustainability 15(2):1022
Wang K (2016) Logistics 4.0 solution-new challenges and opportunities. In: 6th international workshop of advanced manufacturing and automation, pp 68–74. Atlantis Press
Zhang M, Zheng J (2021) Design of warehouse management system based on L company. Indus Eng Innov Manag 4(1):1–14

Use of Digital and Automative Technologies in Greening Supply Chains

The Role of Green Supply Chain Management in Enterprise Development

Łukasz Brzeziński, Magdalena Krystyna Wyrwicka, and Petro Gudz

Abstract Enterprise development is a multi-faceted process that requires continuous adaptation in order to achieve greater efficiency, competitiveness and sustainable growth. Currently, the trends related to green transformation are intensifying, which is also reflected in the functioning of supply chains. Focusing on sustainability and environmental aspects potentially affect a wide range of areas, processes or the philosophy of the organization's operation and can stimulate their development. The aim of the chapter was to analyze the impact of the functioning of the organization in the green supply chain on the development of companies participating in it. Based on the conducted research, it was shown that involvement in green supply chains can in particular affect the technological dimension regarding digital transformation, automation—the use of technology to improve operational efficiency and reduce costs.

Keywords Green supply chain management · Enterprise development · Dimensions of enterprise development · Sustainability

1 Introduction

The development of an enterprise is undoubtedly one of the most complex processes occurring during its operation. At the same time, it is one of the most expected effects of purposeful and organized business activity by stakeholders. This is the highest level of aspiration of the organization (the highest order goal) (Matejun 2015).

Ł. Brzeziński (✉)
Poznan School of Logistics, Poznan, Poland
e-mail: lukasz.brzezinski@wsl.com.pl

M. K. Wyrwicka
Poznan University of Technology, Poznan, Poland
e-mail: magdalena.wyrwicka@put.poznan.pl

P. Gudz
National University Zaporizhzhia Polytechnik, Zaporizhzhia, Ukraine
e-mail: gudz@zntu.eduua

A. Kolinski and M. Adamczak (eds.), *Digitalisation of the Greening Supply Chain*, EcoProduction, https://doi.org/10.1007/978-3-031-88918-9_8

Development should be treated as a comprehensive and long-term process with a strategic basis in the organization, the essence of which is changes. Their subject is primarily the individual elements of the organizational system and the way of implementing individual management functions (Szplit et al. 2002).

Enterprise development can be defined as a continuous process of improving and adapting the organization to achieve greater efficiency, competitiveness, and sustainable growth. This involves various activities aimed at increasing the company's ability to respond to market changes, innovate products and services, and optimize internal operations. Company growth can be analyzed both in quantitative terms, such as revenue increases or market expansion, and qualitative aspects, including enhancing organizational culture, developing employee skills, or increasing customer satisfaction. The central component of this process is management, which integrates different aspects of the company's operations into a cohesive and effective whole, supporting the achievement of long-term objectives.

One the key elements in building an organization's development potential is actions related to supply chain management, which involves the flow of goods, information, and finances associated with a product or service. This flow should be analyzed from the procurement of raw materials to the delivery of the product to its destination. It can occur within a single enterprise or a network of entities. Currently, there are strong trends regarding green transformation within supply chains, which involve the implementation of process monitoring and engaging in pro-ecological and sustainable activities (throughout the entire product life cycle). This approach is identified with green supply chain management.

Such focus on environmental aspects in logistics activities can also generate a range of benefits for organizations. Implementing eco-friendly technological solutions can contribute to reducing greenhouse gas emissions, more efficient resource management, and cost reduction through the application of innovative solutions. By supporting sustainable development, green supply chains not only help achieve financial benefits but also promote broadly understood pro-ecological activities. Therefore, they are not only a key to growth and competitiveness in the market but also an integral component of a responsible business approach, benefitting companies, society, and the environment alike (Lee 2011, pp. 7–8; Roszkowska-Śliż 2011, pp. 57–58; Leszczyńska 2014, p. 83; Baraniecka 2015, p. 246; Nowicka 2015, p. 111).

This emphasizes the significant role of logistics in business development. It highlights that intentional management of processes involving the flow of materials, information, and finances (logistics) ensures efficient order fulfillment, which raises customer satisfaction, and in turn supports sales growth and brand loyalty. By optimizing the supply chain and minimizing inventory levels, companies can significantly reduce operational costs and minimize losses due to delays or delivery errors. Logistics, regarded as an integrating system of material and information flow, not only supports the improvement of customer service levels but also inspires enhancements in forecasting, business planning, and offers numerous indicators of company efficiency and profitability, thereby contributing to its development (Sadowska 2014; Matwiejczuk 2015). However, in the context of analyses concerning the impact of green supply chains on organizational development, there is a research gap involving the omission of significant developmental aspects such as transformations of quantitative parameters (related to scaling operations, increasing market share, product sales), qualitative parameters (related to innovations or management methods), structural and organizational changes, workforce capital (concerning competencies, experience, and knowledge), as well as technological issues (pertaining to automation, digitization) (Qin 2019; Nekmahmud et al. 2020; Ahmad et al. 2022; Afzal and Hanif 2022).

Therefore the aim of the chapter is to analyze the impact of operating within a green supply chain on the development of participating enterprises. The study consists of the following sections: literature review, description of the methodological assumptions of the study, analysis of research results, discussion, and conclusions.

2 Literature Review

Green supply chain management

The concept of the green supply chain emerged in 1989 and has evolved through research on sustainable design, reverse logistics, and waste management. Green supply chains encompass various operational aspects such as design and distribution, aiming to reduce negative environmental impacts and protect natural resources. In the 1990s, researchers conducted detailed analyses of GSCM topics like recycling and green planning (Kelle and Silver 1989; Ashley 1993; Srivastava 2007; Lahkani et al. 2020).

Green supply chain management (GSCM) involves integrating sustainability issues into production, procurement, distribution, sales, marketing, and other areas to enhance productivity and profitability (Srivastava 2007; Green et al. 2012a, b; Grabara et al. 2020). According to Darnall et al. (2008), GSCM practices include ecological actions that provide environmentally friendly products or services and reduce costs in the value chain. GSCM involves implementing and monitoring environmental management programs and creating and controlling practices through activities like reuse (e.g., returnable packaging), reduction (of resource consumption, emissions, or waste), refurbishment (e.g., functionality changes and repairs), recycling (recovering parts or materials), and others that minimize environmental burdens.

GSCM is a coordinated supply chain that considers environmental issues. It also includes internal operational aspects of companies participating in the chain to increase management efficiency of information, material, and capital flows in procurement, production, and distribution, allowing them to meet stakeholder expectations, increase profits, competitiveness, and business resilience (Ahi and Searcy 2013; Kozma 2017; Malkus and Tyranska 2019).

Currently, green supply chain management incorporates sustainable development aspects such as green procurement, production, distribution, and marketing to enhance ecological and social benefits (Nekmahmud et al. 2020).

GSCM encompasses a wide range of activities that in the literature are often aggregated into five dimensions. These dimensions typically include: internal environmental management (IEM), eco-design (ED), green purchasing (GP), cooperation with customers (CWC) including environmental requirements, and investment recovery (IR), which are defined as follows (Laosirihongthong et al. 2013; Khan and Qianli 2017; Gelmez et al. 2024).

- IEM: This aspect is crucial for enhancing corporate performance through the commitment of senior executives and collaboration across different functions (Hsu and Hu 2008). It includes both operational and management practices within the organization.
- ED: This is a comprehensive approach that considers the entire product life cycle, focusing on environmental, health, and safety goals (Thamsatitdej et. al 2017). It integrates ecological considerations into the design and development of environmentally friendly products (Balon 2020). The aim of ED is to minimize the adverse effects of a product on the environment and society during its design phase, without sacrificing other important attributes like performance, usage, or aesthetics (Li and Sarkis 2022).
- GP: This concept involves purchasing raw materials with an emphasis on environmental sustainability and awareness, such as reducing waste, promoting recycling and reuse, conserving resources, and using eco-friendly materials (Sugandini et al. 2020). GP practices are essential for mitigating the environmental impacts associated with production, usage, and recycling processes (Hazaea et. al. 2022).
- CWC: involves businesses collaborating with customers to develop production processes that are environmentally cleaner (Green et al. 2012a, b).
- IR: Investment recovery involves strategically using techniques like redeployment, reselling, and recycling to derive greater value from materials and products in a business. The goal of IR is to transform surplus resources into revenue by selling idle assets, thus avoiding the need for additional equipment or materials, reducing the need for storage, and reallocating unused resources to different parts of the company (Zhu et al. 2008).

With increasing dynamism and popularity of green transformation trends, the impact of activities within green supply chains on business operations is being studied, particularly in the context of sustainable development areas, categorized as environmental, economic, and social (Qin 2019; Li et al. 2025; Knoucha et al. 2025; Kharat et al. 2025).

The Impact of Green Supply Chain Management on Environmental Performance

Johnson (1998) analyzed the relationship between green supply chain management and the balanced scorecard to identify environmental performance indicators within a company's performance evaluation system. It was emphasized that green supply chain management can further enhance a company's financial performance and help achieve business goals by improving environmental efficiency.

Theyel (2000) explored the impact of environmental management practices on ecological innovation and environmental efficiency, showing that these practices can support environmental innovation and efficiency in facilities. Zhu and Sarkis (2004) noted that the most significant outcome of green supply chain management is the improvement of environmental performance. Eltayeb et al. (2011) demonstrated that eco-design can significantly enhance a company's environmental performance.

Additionally, Dai et al. (2014) argue that managers adopting pro-environmental behaviors and implementing green supply chain management can provide companies with a "win–win" situation, improving both environmental performance and financial outcomes. Furthermore, Tseng et al. (2016) found that green supply chain management can reduce negative environmental impacts and ecological risks while eliminating waste and improving eco-efficiency, leading to enhanced environmental performance.

The Impact of Green Supply Chain Management on Economic Performance

Carter et al. (2000) analyzed historical data to investigate the impact of environmentally conscious actions on corporate financial performance. Their research showed a correlation between net profit and product costs. Sarkis (2003) used network analysis to evaluate green supply chain management strategies and indicated that this concept can enhance a company's competitive advantage, improve organizational performance, reduce production costs, and protect the environment.

Corbett and Klassen (2006) assessed that green supply chain management positively affects business performance. Zhu et al. (2010), analyzing 334 manufacturing companies from Dalian, Suzhou, and Tianjin, demonstrated a link between pro-environmental collaboration in the supply chain (green purchasing, cooperation with customers) and circular economy practices (internal environmental management, eco-design, investment recovery) and their performance. This suggests that a higher level of green purchasing and customer collaboration leads to better internal environmental management, eco-design, and investment recovery, resulting in improved environmental and economic performance.

Green et al. (2012a, b) applied structural equation modeling to analyze data from 159 manufacturing enterprises. The study results showed that the adoption of green supply chain management technologies in manufacturing companies positively influenced environmental and economic performance, leading to increased operational efficiency, which in turn improved organizational outcomes.

The Impact of Green Supply Chain Management on Social Performance

Analysis the social context of business operations derives from the theory of corporate social responsibility (CSR) and pertains to the actions undertaken by companies to prevent, mitigate, or correct negative external impacts of their activities, as well as broader social issues that do not necessarily arise directly from the organizations' operations. Social outcomes are evaluative results of activities related to corporate social responsibility. Social issues such as product safety and environmental pollution have garnered significant interest from researchers in recent years, and social outcomes are also considered a key factor in green supply chains (Qin 2019).

Freeman and Gilbert (1998) indicated that, in addition to striving for financial results for shareholders, companies should also assume social responsibility and seek benefits for other stakeholders. Furthermore, based on Preuss's (2000) research, it has been confirmed that environmental protection activities undertaken within a company can be transferred to suppliers through purchasing functions, creating a domino effect that leads to rapid and profound changes in overall social outcomes. Eiadat e tal. (2008) demonstrated that implementing green innovation strategies can help companies gain a good social reputation, allowing them to achieve first-mover advantage in environmental protection and thus enhance their social performance. Additionally, Chiu and Sharfman (2011) argue that corporate social outcomes result from actions promoting social responsibility, and environmentally responsible behaviors can enhance companies' social performance.

Based the analyses presented, it is possible to indicate the potential impact of green supply chains on the development of enterprises in the described key areas: environmental, economic, and social. However, the phenomenon of development is much broader and also encompasses other aspects of organizational functioning, which will be explored in the subsequent sections of the article.

3 Research Methodology

The aim of the chapter is to describe the methodological assumptions and organization of the study. At the same time, based on a review of the subject literature, potential areas of dependence between GSCM and organizational development were identified.

The impact of green supply chain management on enterprise development will be analyzed based on the dimensions of enterprise development potential relating to quantitative growth (expansion), qualitative changes, structure, human resources, technology, finances and ecological and social aspects (Matwiejczuk 2019; Strabyła 2014; Świetlińska 2014; Brzeziński 2024).

Various methods were used in the research depending on the phase (preparation, implementation, processing of results)—Fig. 1.

Below is a description and an indication of their use in individual stages of the study.

Research Preparation Phase

The aim of the study is to analyze the impact of operating within a green supply chain on the development of participating enterprises.

The authors formulated the following research problems:

- Q1: What is the level of impact of green supply chain management on the quantitative growth of the organization (based on expert opinions)?
- Q2: What is the level of impact of green supply chain management on the qualitative development of the enterprise (based on expert opinions)?

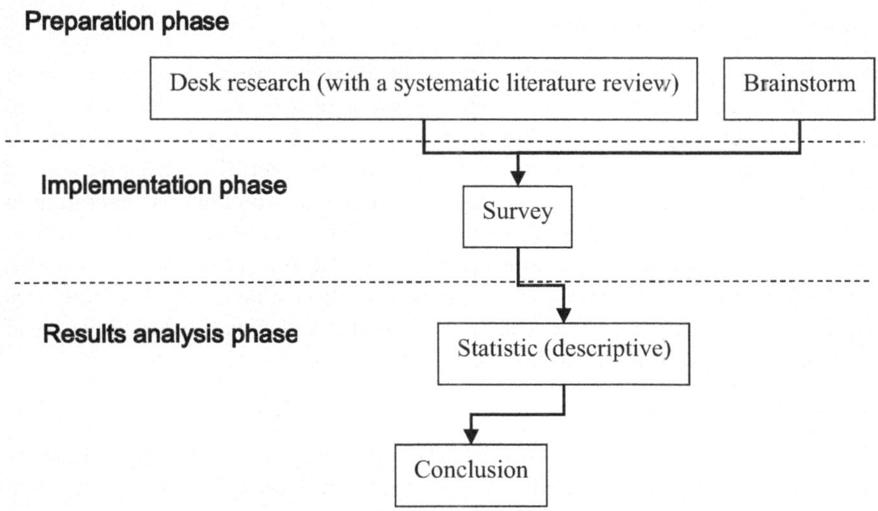

Fig. 1 The scheme of the author's research, own study

- Q3: What is the level of impact of green supply chain management on the structural development of the enterprise (based on expert opinions)?
- Q4: What is the level of impact of green supply chain management on the development of the human resources of the enterprise (based on expert opinions)?
- Q5: What is the level of impact of green supply chain management on the technological development of the enterprise (based on expert opinions)?
- Q6 What is the level of impact of green supply chain management on the financial development of the enterprise (based on expert opinions)?
- Q7: What is the level of impact of green supply chain management on the ecological and social development of the enterprise (based on expert opinions)?

Desk research entails analyzing accessible data sources, focusing on compiling, cross-verifying, and processing them. This analytical process forms the foundation for drawing conclusions regarding the researched issue (Makowska 2013; Bednarowska 2015).

Brainstorming is an activating work method that is based on generating solutions to given problems spontaneously. It involves working in a group, the aim of which is to generate ideas for finding the causes of the problem, solutions and selecting the best options (Gołaś and Mazur 2010).

Both methods were used to develop an survey form containing questions and suggestions regarding the assessed dimensions relating to the enterprise's development potential relating to quantitative growth (expansion), qualitative changes, structure, human resources, technology, finances and ecological and social aspects (Matwiejczuk 2019; Strabyła 2014; Świetlińska 2014; Brzeziński 2024)

Research implementation phase

The survey is a commonly used research method that involves collecting data using standardized questions addressed to a specific group of respondents (Samuelson and Marks 2009). A purposive sampling technique was used—the survey questionnaire was sent electronically to owners and representatives of enterprises (from the sectors of: transport and logistics, construction, e-commerce and production) operating in Poland, which operate within green supply chains. The condition for participation in the survey was the functioning of the organisation within green supply chains—this was an element included in the survey and the invitation to the study. The study was conducted in October–November 2024. 43 respondents took part in the study (a total of 193 surveys were sent to enterprises).

Results analysis phase

The analysis of the collected data was based on statistical analysis including descriptive measures such as arithmetic mean, dominant, minimum value, maximum value and range (Ręklewski 2020):

- Arithmetic average is the sum of the value of variable of all units of the studied population divided by the number of units of the population.
- The dominant is the value of the variable that is the most frequent (dominant, typical) in the studied community. The dominant is called a modal value or mode.
- The minimum value of recorded responses in accordance with the adopted scale.
- The maximum value of the recorded responses in accordance with the adopted scale.
- The range is the difference between the maximum and minimum.

4 Results

The survey was conducted on a group of 43 respondents. The largest percentage were business owners (20.83%), followed by managers (29.17%), people in management positions (14.58%), board members (10.42%) and the remaining 4.17% were directors (Fig. 2).

In terms of enterprise size, the largest percentage of survey participants represented small enterprises (from 10 to 49 employees—45.83%) and medium-sized enterprises (from 50 to 250 employees—29.17%). In addition, large enterprises (employing more than 249 employees) accounted for 8.33%, and the remaining 6.25% were microenterprises (fewer than 10 employees)—Fig. 3.

The analysed companies represented the following industries: transport and logistics (41.86%), e-commerce (27.91%), production (20.93%) and construction (9.30%)—Fig. 4.

The impact of green supply chain management on enterprise development has been analyzed based on the dimensions of enterprise development potential relating to quantitative growth (expansion), qualitative changes, structure, human resources,

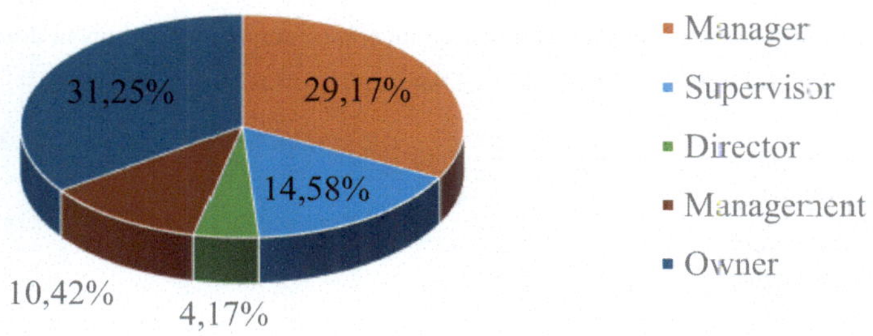

Fig. 2 Respondent's job position, own study

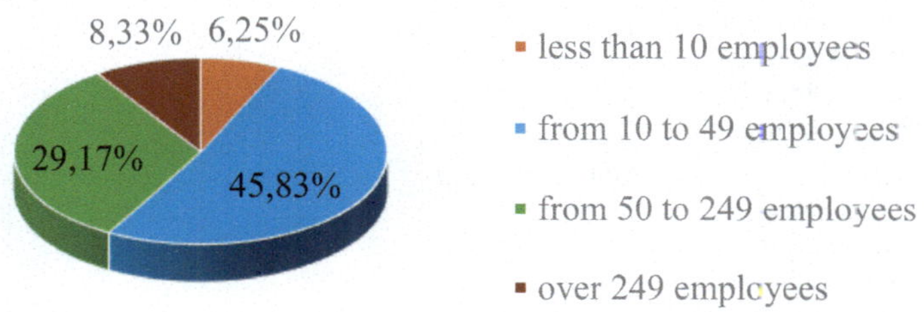

Fig. 3 Size of the company operating in the green supply chain, own study

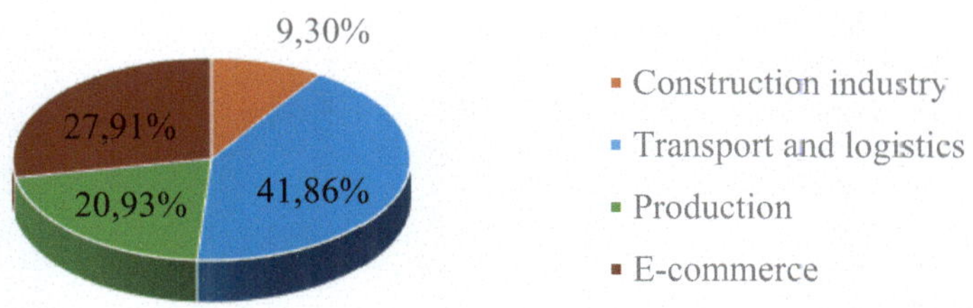

Fig. 4 Green supply chain industry, own study

technology, finances and ecological and social aspects (Matwiejczuk 2019; Strabyła 2014; Świetlińska 2014; Brzeziński 2024)—Table 1.

The data presented in the table reflects the impacts of operating within green supply chains on enterprise development across various dimensions. Overall, the findings indicate a moderate to strong commitment among businesses to pursue both quantitative and qualitative growth, highlighting the importance of sustainability in their strategic initiatives.

In the dimension of **quantitative growth**, the average score of 3.24 suggests that enterprises recognize the significance of expansion but may not prioritize it to the highest extent. Notably, scaling up production received a commendable score of 3.86, indicating a strong inclination toward investing in eco-efficient production capabilities. This demonstrates that companies are gearing up to enhance their operational

Table 1 The impact of green supply chain management on enterprise development dimensions

Development dimensions		Average	Dominant	Maximal value	Minimal value	Range
1	*Quantitative growth (expansion)*	*3,24*	*3,50*	*5*	*1*	*4*
1.1.	Scaling up production: increasing production capacity by investing in eco-efficient production lines, machinery, or hiring skilled workers knowledgeable in green practices	3,86	4,00	5	2	3
1.2.	Market development: expanding into new geographic markets, both domestic and international, with a focus on eco-friendly products to increase the customer base	3,00	3,00	4	2	2
1.3.	Sales expansion: intensifying marketing efforts to highlight sustainable practices, thereby increasing sales volume	3,35	4,00	4	2	2
1.4.	Product expansion: launching new eco-friendly products or services to meet a wider range of customer needs	2,74	3,00	4	1	3
2	*Qualitative development*	*3,51*	*3,00*	*5*	*2*	*3*
2.1.	Product innovation: introducing environmentally innovative products or services that offer unique sustainable benefits	3,05	3,00	5	2	3

(continued)

Table 1 (continued)

Development dimensions		Average	Dominant	Maximal value	Minimal value	Range
2.2.	Quality management: enhancing production and operational processes to improve product quality while reducing environmental impact	4,14	4,00	5	3	2
2.3.	Brand reinforcement: strengthening brand position by improving the image of environmental responsibility and increasing customer loyalty	3,35	3,00	5	2	3
3	*Structural development*	*3,09*	*3,00*	*5*	*2*	*3*
3.1.	Organizational restructuring: adapting organizational structures to support green initiatives, such as creating sustainability departments	2,77	3,00	4	2	2
3.2.	Mergers and acquisitions: partnering with or acquiring companies with strong environmental practices to enhance resources and know-how	2,56	2,00	4	2	2

(continued)

Table 1 (continued)

Development dimensions		Average	Dominant	Maximal value	Minimal value	Range
3.3.	Change management: implementing flexible management strategies to accommodate sustainable business practices	3,95	4,00	5	2	3
4	*Human resource development*	*3,55*	*4,00*	*5*	*1*	*4*
4.1.	Employee training and development: investing in environmental training programs to enhance employee efficiency and innovation in sustainability	3,77	4,00	5	2	3
4.2.	Organizational culture: cultivating a culture that values sustainability, cooperation, and innovation	3,72	4,00	5	2	3
4.3.	Talent management: attracting and retaining employees committed to sustainability initiatives	3,16	3,00	5	1	4
5	*Technological development*	*4,35*	*5,00*	*5*	*2*	*3*
5.1.	Automation: implementing eco-friendly automation technologies to increase efficiency and reduce costs	4,35	4,00	5	3	2

(continued)

Table 1 (continued)

Development dimensions		Average	Dominant	Maximal value	Minimal value	Range
5.2.	Digital transformation: utilizing digital technologies to enhance sustainable practices across management, sales, and marketing	4,47	5,00	5	2	3
5.3.	Research and development (R&D) investments: funding research into green technologies for competitive advantages in sustainability	4,23	5,00	5	2	3
6	*Financial development*	*3,42*	*3,00*	*5*	*2*	*3*
6.1.	Raising capital: securing funds through green bonds or eco-investments	3,19	3,00	4	2	2
6.2.	Cost management: reducing operational costs through sustainable practices and resource optimization	3,77	4,00	5	2	3
6.3.	Investment strategies: focusing on green technologies or sustainable market expansions	3,30	3,00	4	2	2
7	*Ecological and social development*	*4,12*	*4,00*	*5*	*2*	*3*
7.1.	Sustainable development: implementing practices that minimize environmental impact and foster long-term ecological balance	4,56	5,00	5	3	2

(continued)

Table 1 (continued)

Development dimensions		Average	Dominant	Maximal value	Minimal value	Range
7.2.	Corporate social responsibility (CSR): engaging in environmentally and socially responsible activities to improve public perception	3,72	4,00	5	2	3
7.3.	Risk management: developing strategies to mitigate environmental and social risks associated with business operations	4,07	4,00	5	2	3

Source own study

capacity and are focused on incorporating green practices, which is essential for long-term sustainability. However, the lowest score was recorded in product expansion at 2.74. This reveals a lack of emphasis on diversifying their eco-friendly product offerings, suggesting that while firms are scaling their operations, they may not be fully capitalizing on innovation in product lines.

Moving on to **qualitative development**, which achieved an average of 3.51, the firms exhibit a robust focus on improving product quality and reinforcing brand identity. The particularly high score for quality management at 4.14 indicates that businesses are making substantial efforts to enhance operational processes while simultaneously reducing their environmental footprint. However, the score for product innovation at 3.05 suggests a need for improved efforts in introducing innovative products that offer unique sustainable benefits, emphasizing a critical area of growth yet to be fully realized.

In **structural development**, the average score of 3.09 indicates a moderate commitment to reshaping organizational structures to better support green initiatives. The highest individual score here is for change management at 3.95, which emphasizes that companies are willing to adapt their management strategies to accommodate sustainability. However, the lower scores for mergers and acquisitions at 2.56 imply that companies may not be actively seeking strategic partnerships with organizations that have strong environmental practices, potentially limiting their access to valuable resources and knowledge.

The dimension of **human resource development** registered a solid average of 3.55, suggesting that organizations are placing importance on their workforce in the context of sustainability. High scores for both employee training and development at

3.77 and organizational culture at 3.72 indicate a recognition of the need to cultivate a skilled workforce and an organizational ethos that values sustainable practices. In contrast, the score for talent management at 3.16 points to challenges in attracting and retaining employees who are deeply committed to sustainability initiatives.

Technological development stands out as the strongest area, achieving an average score of 4.35. This suggests that enterprises are heavily investing in technology to support their sustainable efforts. The scores for digital transformation (4.47) and automation (4.35) indicate a proactive approach towards leveraging technology to enhance operational efficiency and reduce costs, foundational elements for a successful green transition.

The average score for **financial development** at 3.42 shows that enterprises are applying sustainable practices to enhance their financial strategies. The relatively low score for raising capital at 3.19 underscores potential difficulties in securing funds specifically for green initiatives, despite a strong focus on cost management (3.77) which highlights operational efficiency.

Finally, the dimension of **ecological and social development** scores the highest average at 4.12, reflecting a strong commitment to sustainable development practices. Notably, sustainable development achieves an impressive 4.56, emphasizing that organizations prioritize implementing practices that minimize their environmental footprint. The scores for corporate social responsibility at 3.72 and risk management at 4.07 reinforce the notion that companies are not just focused on their own sustainability, but are also engaged in socially responsible activities to enhance public perception and mitigate risks associated with their operations.

The visualization of the results is presented in Fig. 5.

In summary, the analysis reveals that companies operating within green supply chains recognize the need for both quantitative and qualitative growth. They show

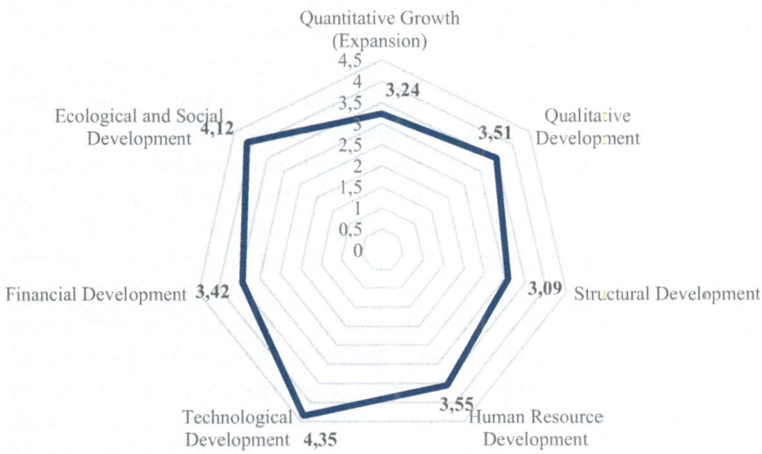

Fig. 5 Visualization of results in relation to the impact of green supply chain management on the analyzed dimensions of enterprise development, own study

commendable efforts in implementing technologies that support sustainability, but there is still room for improvement in areas such as product innovation and proactive structural changes. These findings suggest that enhancing innovation and fostering strategic partnerships could further support the overall development and effectiveness of enterprises within the green economy.

The study highlights the importance of Green Supply Chain Management (GSCM) practices, especially eco-design, in enhancing environmental, economic, and social performance. This reinforces existing literature showing that sustainable products, through recyclability and reduced emissions, positively impact market perception and value.

Practically, the survey underscores the commitment of businesses to invest in sustainable technologies, as seen in high scores for technological development and digital transformation. Despite this, there is room for growth in product innovation and strategic partnerships. Businesses can leverage these insights to enhance their competitive edge in the green economy by focusing on diversifying their eco-friendly product lines and strengthening collaboration with environmentally conscious partners.

5 Discussion

It should be noted that the impact of involvement in green supply chains is referred to in the literature on the subject, especially in the dimensions of sustainable development—i.e. economic, social and environmental. Change or improvement of functioning in individual areas may affect the development of the organization, but such a comprehensive approach to changes is not one of the popular directions of analysis and is difficult to find in the literature. However, based on selected analyses, several important developmental aspects can be distinguished.

Eco-design stands out as the most impactful GSCM practice, significantly enhancing economic, social, and environmental performance. This aligns with previous research by Khan et al. (2022) and Sarwar et al. (2021), highlighting that products with eco-design are preferred due to their recyclability, reduced energy consumption, and lower emissions, which in turn improve their market image and resale value. Green purchasing also plays a crucial role, positively affecting social and environmental outcomes, as noted by Ahmad et al. (2022) and Zailani et al. (2015).

Contrary to expectations, green production does not correlate with economic or social performance, diverging from other studies like those by Afzal and Hanif (2022) and Sarwar et al. (2021). This discrepancy may arise from varying perspectives among respondents and the current maturity level of these organizations in adopting green production practices. However, it does positively impact environmental performance, corroborated by Ahmad et al. (2022) and Govindan et al. (2015), indicating an awareness of environmental issues and proactive protection measures. Cooperation with customers shows a positive link with social performance, supporting findings by Sarwar et al. (2021) about green production enhancing social outcomes.

However, it does not appear to influence economic or environmental performance, possibly due to a lack of strategic focus on customer relations within these enterprises. Lastly, green logistics, while not connected with economic or social performance, does positively influence environmental performance, consistent with Kerdpitak (2021).

6 Conclusions and Further Research

Conducted analysis of the impact of green supply chain management on enterprise development, highlighting its complexity and significance in the context of modern business operations. It emphasizes that enterprise development is a multifaceted process requiring continuous adaptation to achieve greater efficiency, competitiveness, and sustainable growth. The findings suggest that businesses engaged in green supply chains are demonstrating a moderate to strong commitment to both quantitative and qualitative growth, particularly focusing on enhancing production capabilities and quality management.

While the investment in eco-efficient production and a commitment to quality is evident, there is a notable gap in product innovation and diversification, indicating that companies may be missing opportunities to expand their market presence through sustainable product offerings. The data also highlights the importance of organizational structures that support green initiatives, as well as the need for effective change management strategies.

Furthermore, the analysis demonstrates that organizations recognize the critical role of human resource development, particularly in training and cultivating a culture that promotes sustainability. However, challenges remain in attracting and retaining talent fully committed to sustainability initiatives.

The conducted study has some limitations. In particular, the selected research group is not representative, therefore the results cannot be generalized to the general population—entities operating in green supply chains in Poland. It should be emphasized that estimating the number of companies that operate in this way is problematic, statistics in this area are not kept. However, the results may constitute an added value due to their exploratory nature. In the future, further studies may refer to individual industries, or estimate the number of entities in a given industry that operate within green supply chains.

References

Afzal N, Hanif A (2022) The impact of green supply chain management practices on firm performance: evidence from manufacturing industry. Global Bus Rev, 1.https://doi.org/10.1177/09721509221125576

Ahi P, Searcy C (2013) A comparative literature analysis of definitions for green and sustainable supply chain management. J Clean Prod 52:329–341

Ahmad A, Ikram A, Rehan MF, Ahmad A (2022) Going green: impact of green supply chain management practices on sustainability performance. Front Psychol 13:973676. https://doi.org/10.3389/fpsyg.2022.973676

Ashley S (1993) Designing for the environment. Mech Eng CIME 115(3):52–56

Balon V (2020) Green supply chain management: Pressures, practices, and performance-an integrative literature review. Bus Strategy Dev 3:226–244

Baraniecka A (2015) Rozwój ekologistycznych łańcuchów dostaw jako skutek kryzysów: ekonomicznego i środowiskowego. Prace Naukowe Uniwersytetu Ekonomicznego we Wrocławiu, 383

Bednarowska Z (2015) Desk research—Wykorzystanie potencjału danych zastanych w prowadzeniu badań marketingowych i społecznych. Marketing i Rynek 7:18–26

Brzeziński Ł (2024) The importance of logistics in the development of an enterprise. 24rd international scientific conference business logistics in modern management October 3–4, 2024—Osijek, Croatia

Carter CR, Kale R, Grimm CM (2000) Environmental purchasing and firm performance: an empirical investigation. Transp Res Part E Logist Transp Rev 36:219–228. https://doi.org/10.1016/S1366-5545(99)00034-4

Chiu SC, Sharfman M (2011) Legitimacy, visibility and the antecedents of corporate social performance: an investigation of the instrumental perspective. J Manag 37:1558–1585. https://doi.org/10.1177/0149206309347958

Corbett CJ, Klassen RD (2006) Extending the horizons: environmental excellence as key to improving operations. Manuf Serv Oper Manag 8:5–22. https://doi.org/10.1287/msom.1060.0095

Dai J, Montabon FL, Cantor DE (2014) Linking rival and stakeholder pressure to green supply management: mediating role of top management support. Transp Res Part E Logist Transp Rev 71:173–187. https://doi.org/10.1016/j.tre.2014.09.002

Darnall N, Jolley GJ, Handfield R (2008) Environmental management systems and green supply chain management: complements for sustainability? Bus Strateg Environ 17(1):30–45

Eiadat Y, Kelly A, Roche F, Eyadat H (2008) Green and competitive? An empirical test of the mediating role of environmental innovation strategy. J World Bus 43:131–145. https://doi.org/10.1016/j.jwb.2007.11.012

Eltayeb TK, Zailani S, Ramayah T (2011) Green supply chain initiatives among certified companies in Malaysia and environmental sustainability: investigating the outcomes. Resour Conserv Recycl 55:495–506. https://doi.org/10.1016/j.resconrec.2010.09.003

Freeman RE, Gilbert DR (1988) Corporate strategy and the search for ethics, Vol 1, Prentice Hall Englewood Cliffs, NJ

Gelmez E, Özceylan E, Mrugalska B (2024) The impact of green supply chain management on green innovation, environmental performance, and competitive advantage. Sustainability 16(22):9757. https://doi.org/10.3390/su16229757

Gołaś H, Mazur A (2010) Zasady, metody i techniki wykorzystywane w zarządzaniu jakością. Poznań: Wydawnictwo Politechniki Poznańskiej

Govindan K, Diabat A, Shankar KM (2015) Analyzing the drivers of green manufacturing with fuzzy approach. J Clean Prod 96:182–193. https://doi.org/10.1016/j.jclepro.2014.02.054

Grabara J, Dabylova M, Alibekova G (2020) Impact of legal standards on logistics management in the context of sustainable development. Acta Logistica 7(1):31–37

Green KW Jr, Zelbst PJ, Meacham J, Bhadauria VS (2012) Green supply chain management practices: impact on performance. Supply Chain Manag Int J 17:290–305. https://doi.org/10.1108/13598541211227126

Green KW, Zelbst PJ, Meacham J, Bhadauria VS (2012) Green supply chain management practices: impact on performance. Supply Chain Manag Int J 17(3):290–305

Hazaea SA, Al-Matari EM, Zedan K, Khatib SF, Zhu J, Al Amosh H (2022) Green purchasing: past, present and future. Sustainability 14:5008

Hsu CW, Hu AH (2008) Green supply chain management in the electronic industry. Int J Environ Sci Technol 5:205–216

Johnson SD (1998) Identification and selection of environmental performance indicators: application of the balanced scorecard approach. Corp Environ Strateg 5:34–41. https://doi.org/10.1016/S1066-7938(00)80079-2

Kelle P, Silver EA (1989) Forecasting the returns of reusable containers. J Oper Manag 8(1):17–35

Kerdpitak C (2021). he role of green logistics and supply chain management of recycling waste management on green environment of thailand. Ilkogretim Online 20(4):1843–1855. Filip A., CSR i zrównoważony rozwój. Available online: https://v4l.pl/csr-i-zrownowazony-rozwoj-na-europejskich-szlakach-logistycznych/. Access 5 Dec 2024

Khan MT, Idrees MD, Rauf M, Sami A, Ansari A, Jamil A (2022) Green supply chain management practices' impact on operational performance with the mediation of technological innovation. Sustainability 14(6):3362. https://doi.org/10.3390/su14063362

Khan SAR, Qianli D (2017) Impact of green supply chain management practices on firms' performance: an empirical study from the perspective of Pakistan. Environ Sci Pollut Res 24:16829–16844

Kharat MG, Kapoor S, Parhi S, Kharat MG, Pandey S (2025) Operationalizing sustainability in pharmaceuticals: green supply chain metrics for circular economy. Sustain Futures 9. https://doi.org/10.1016/j.sftr.2024.100413

Knoucha M, Lefkirb H, Alqahtani HA (2025) The effects of environmental sustainability orientation on the sustainable performance, the mediating role of green supply chain management and moderating role of environmental collaboration: Evidence from Algeria. Decision Sci Lett 14:145–154. https://doi.org/10.5267/j.dsl.2024.10.006

Kozma T (2017) Cooperation in the supply chain network. Forum Scientiae Oeconomia 5(3):45–58

Lahkani MJ, Wang S, Urbański M, Egorova M (2020) Sustainable B2B E-commerce and blockchain-based supply chain finance. Sustainability 12(10):3968

Laosirihongthong T, Adebanjo D, Choon Tan K (2013) Green supply chain management practices and performance. Ind Manag Data Syst 113:1088–1109

Lee H (2011) Nie majsterkuj przy swoim łańcuchu dostaw, zaprojektuj go od nowa. Harv Bus Rev Polska, 3

Leszczyńska A (2014). Zastosowanie wskaźników rozwoju zrównoważonego w łańcuchu dostaw. Optimum Studia Ekonomiczne, 4

Li D, Zhang Z, Xu R (2025) The impact of environmental regulation on green image management of supply chain: evidence from China. Financ Res Lett 74. https://doi.org/10.1016/j.frl.2024.106723

Li J, Sarkis J (2022) Product eco-design practice in green supply chain management: a China-global examination of research. Nankai Bus Rev Int 13:124–153

Makowska M (2013) Analiza Danych Zastanych. Przewodnik dla Studentów. Wydawnictwo Naukowe Scholar, Warszawa

Malkus T, Tyranska M (2019) Reaction to risk in logistics cooperation-results of empirical research. Acta Logistica 6(3):77–84

Matejun M (2015) Absorpcja wsparcia w zarządzaniu rozwojem mikro, małych i średnich przedsiębiorstw—podejście strategiczne, Zeszyty Naukowe Nr 1194, Rozprawy Naukowe Z. 483. Łódź: Wydawnictwo Politechniki Łódzkiej

Matwiejczuk R (2015) Z badan nad logistycznymi determinantami zarządzania przedsiębiorstwem. Studia Ekonomiczne. Zeszyty Naukowe Uniwersytetu Ekonomicznego w Katowicach, 251

Matwiejczuk R (2019) Zarządcze aspekty logistyki jako determinanty rozwoju potencjałów strategicznych przedsiębiorstwa. Studia i Prace, Kolegium Zarządzania i Finansów, Zeszyt Naukowy 175:105–114

Nekmahmud M, Rahman S, Sobhani F, Olejniczak-Szuster K, Fekete-Farkas M (2020) A systematic literature review on development of green supply chain management. Polish J Manag Studies 22(1):351–370. https://doi.org/10.17512/pjms.2020.22.1.23

Nowicka K (2015) Innowacje w logistyce miejskiej—ITS jako usługa. Prace Naukowe Uniwersytetu Ekonomicznego we Wrocławiu, 383

Preuss L (2000) Should you buy your customer's values? On the transfer of moral values in industrial purchasing. Int J Value Based Manag 13:141–158. https://doi.org/10.1023/A:1007798200832

Qin ZY (2019) A literature review of the impact of green supply chain management on firm performance. J Serv Sci Manag 12:872–879

Ręklewski M (2020) Statystyka Opisowa. Włocławek: Państwowa Uczelnia Zawodowa we Włocławku

Roszkowska-Śliż M (2011) Budowanie relacji z dostawcami w świetle koncepcji CSR, [w:] N. Ćwik (eds.), Wspólna odpowiedzialność. Rola dostaw i zakupów. Warszawa: Forum Odpowiedzialnego Biznesu

Sadowska B (2014) Rola i miejsce logistyki w systemie zarządzania przedsiębiorstwem. ZS WSH Zarządzanie 1:179–186

Sarkis J (2003) A strategic decision framework for green supply chain management. J Clean Prod 11:397–409. https://doi.org/10.1016/S0959-6526(02)00062-8

Sarwar A, Zafar A, Hamza MA, Qadir A (2021) The effect of green supply chain practices on firm sustainability performance: evidence from Pakistan. Uncertain Supply Chain Manag 9(1):31–38. https://doi.org/10.5267/j.uscm.2020.12.004

Srivastava SK (2007) Green supply- chain management: a state- of- the- art literature review. Int J Manag Rev 9(1):53–80

Strabyła A (2014) Analiza zdolności rozwojowej przedsiębiorstwa w kontekście innowacyjności. Studia Ekonomiczne 183:2

Sugandini D, Susilowati C, Siswanti Y, Syafri W (2020) Green supply management and green marketing strategy on green purchase intention: SMEs cases. J Indus Eng Manag 13:79–92

Świetlińska M (2014) Zrównoważony łańcuch żywności—studium przypadku. Prace Naukowe Akademii im. Jana Długosza w Częstochowie, 8

Szplit A, Fudaliński J, Markiewicz P, Smutek H (2002) Strategie rozwoju organizacji. Kraków: Antykwa

Thamsatitdej P, Boon-Itt S, Samaranayake P, Wannakarn M, Laosirihongthong T (2017) Eco-design practices towards sustainable supply chain management: interpretive structural modelling (ISM) approach. Int J Sustain Eng 10:326–337

Theyel G (2000) Management practices for environmental innovation and performance. Int J Oper Prod Manag 20:249–266. https://doi.org/10.1108/01443570010304288

Tseng ML, Tan K, Chiu AS (2016) Identifying the competitive determinants of firms' green supply chain capabilities under uncertainty. Clean Technol Environ Policy 18:1247–1262. https://doi.org/10.1007/s10098-015-1064-0

William FS, Marks SG (2009). Ekonomia menedżerska. Warszawa: Polskie Wydawnictwo Ekonomiczne

Zailani S, Govindan K, Iranmanesh M, Shaharudin MR, Sia Chong Y (2015) Green innovation adoption in automotive supply chain: the Malaysian case. J Clean Prod 108:1115–1122. https://doi.org/10.1016/j.jclepro.2015.06.039

Zhu Q, Sarkis J (2004) Relationships between operational practices and performance among early adopters of green supply chain management practices in Chinese manufacturing enterprises. J Oper Manag 22:265–289. https://doi.org/10.1016/j.jom.2004.01.005

Zhu Q, Geng Y, Lai KH (2010) Circular economy practices among Chinese manufacturers varying in environmental-oriented supply chain cooperation and the performance implications. J Environ Manage 91:1324–1331. https://doi.org/10.1016/j.jenvman.2010.02.013

Zhu Q, Sarkis J, Cordeiro JJ, Lai KH (2008) Firm-level correlates of emergent green supply chain management practices in the Chinese context. Omega 36:577–591

Logistics on Loop: Innovative Supply "Chains" of the Circular Economy

Jakob Grubmüller and Sascha Düerkop

Abstract Supply Chains are traditionally linear systems, in which raw materials go through a process of value creation to become a product that is sold to the consumer and finally reaches the end of the chain. After use, products usually transform to solid waste, to be burned or used as landfill on a large scale. Emerging calls for a Circular Economy (CE), in which products are designed to be constantly recycled and repaired—and in which waste of resources and emissions are minimized—could modify linear supply chains in the future. Many policy makers and industry leaders strive towards this goal. The anticipated sustainable resource usage of CE would drastically change logistics operations and services. This work takes a closer look at current tools and practices in logistics, focusing on their compatibility with this potential major change of economic activity. It will further analyze the implications, which a shift from linear supply chains to circular supply chains would have on logistics operations in general. The paper aims to improve awareness for the necessary adjustments in logistics and therefore enhance readiness of distribution systems. Related recommendations for stakeholder action and areas for further research are identified.

Keywords Circular economy · Green logistics · Emission-free · Reverse logistics · Waste reduction

1 Introduction

An increasing amount of evidence in a range of relevant scientific fields indicates that designing economic activity in a sustainable manner becomes more pressing to reach milestones against global warming. The research consensus at this point in time can

J. Grubmüller (✉)
Department of Business, Fulda University of Applied Sciences, Fulda, Germany
e-mail: Jakob.Grubmueller@w.hs-fulda.de

S. Düerkop
Fraunhofer Institute for Technological Trend Analysis, Euskirchen, Germany
e-mail: Sascha.Dueerkop@int.fraunhofer.de

A. Kolinski and M. Adamczak (eds.), *Digitalisation of the Greening Supply Chain*, EcoProduction, https://doi.org/10.1007/978-3-031-88918-9_9

be described as promoting measures to limit expected damages, as an atmospheric warming of up to one point five degrees Celsius can almost certainly not be avoided any longer (Lamboll et al. 2023). Extreme warming of three degrees Celsius or more is considered by some as incompatible with human civilization in today's definition. A lack of geographic connection of cause and effect and a generational delay in the most violent impacts still hinder global coordination and comprehensive efforts in this matter. Furthermore, the fallout from a changing climate is expected and observed to disproportionately affect those areas, which are least able to cope without external support (Mendelson et al. 2006).

At the same time, a persisting reliance on the extensive extraction of non-renewable resources causes predictable scarcity. Especially the use of fossil fuels, itself the major contributor to climate change, is limited to the reserves of these resources. The goal to minimize global warming requires large quantities of fossil resources to remain in the ground (Welsby et al. 2021). Likewise, an efficient use of electric alternatives, powered by renewable energies, requires the use of finite elements, e.g. rare earths. While these are no longer considered rare, their extraction will become more challenging with time, as the global demand for power storage is expected to increase exponentially. A meaningful measurement of resource overuse is the so-called Earth Overshoot Day, the day of each year at which all sustainably available resources have been used up and unsustainable overuse begins. This marker moved from the end of December to July in the last 50 years. These existing and anticipated scarcities are contrasted by increasing amounts of non-recycled solid waste (Lehmacher and Bödecker 2023). The burning of these vast quantities of refuse, its storage below ground or disposal of non-organic matter into nature as landfill showcase failure to manage economies in an efficient and sustainable manner: only five percent of global waste value is currently maintained in a meaningful way (Fennemann et al. 2018). Recovery of energy through incineration is small consolation, as there are considerable side-effects to this practice, e.g. toxic ashes and pollution. The omnipresence of microscopic particles of plastic in water sources and therefore all organisms indicates the drastic necessity for global action on waste. It is usually agreed upon that avoidance of waste is far more efficient in that endeavor than all disposal and recycling efforts. A comprehensive approach in this would be able to address both the unsustainable transformation of valuable resources into harmful waste and the related generation of greenhouse gas emissions, which drives the accelerating behemoth of climate change.

To tackle these seemingly insurmountable challenges of our time, comprehensive approaches to change provide the greatest potential in reducing harmful consequences of economic activity. One increasingly popular proposal of such a comprehensive action is the move towards "circularity" in economic activity, broadly summarized in the term "Circular Economy" (CE). In general, this concept aims to keep as many resources as possible in use, in contrast to consuming and disposing of the valuable means. To achieve this goal, CE concepts suggest designing the sourcing, production and disposal—but also distribution—in all economic areas around the effort to preserve and reuse resources or products as long as possible and eliminating the concept of useless or even toxic waste or emissions (Morseletto 2020). Current research suggests the potential of designing economic activity in such a way (Fennemann et al. 2018; Lehmacher and Bödecker 2023). A common visualization of CE is the treatment of planet earth as an enormous space station. As all aspects of the latter are designed carefully to extend usability, satisfy multiple needs in their lifecycle and produce practically no waste, so does a circular approach to economic activity treat the finite resources of earth (Haas et al. 2020). This extremely ambitious concept of reorganization towards sustainable use of resources knows many varieties in scale and extent. It can be proposed globally, nationally or within a single organization. All of these options for CE share that previous modes of economic activity have to be restructured comprehensively.

Logistics pose a large part of current economies. It contributes to economic development through its connecting role in production, distribution and disposal. At the same time, transportation directly emits a significant share of total greenhouse gas emissions through its predominant use of fossil fuels (Ritchie 2020; Sosnowski 2022). Overarching all economic activities, the management of supply chains includes both the benefits and the downsides of the linear economy. This straight relationship from source to sink is not a viable option in a circular economy, where the sale to the consumer (in CE rather called 'the user') is not the final step. Accordingly, in the field of supply chain management (SCM), all tools, models and concepts require a stern examination in regard to their compatibility with CE. More specifically, the distribution of goods via logistical means is suspected by the authors to be ill-equipped for this comprehensive change. For instance, the crucial logistical aspect of the so-called 'last mile', the transportation of goods to its final destination, becomes obsolete as a concept and needs reevaluation and restructuring, depending on the circular attributes of the distributed goods. The final destination itself is to be eliminated from logistics, as the supply "chain" will not end with the respective user of goods. The quotation marks in this paper's title therefore suggest that the metaphorical chain (already rather a net) is no longer a fitting nomenclature, as supply is to be increasingly moved in circular fashion.

It becomes evident that not only product designs will need revision to acquire circular properties, logistics practices will have to adapt as well. A comprehensive (global or national) move towards CE would have to include SCM and logistics as a vital part of its potential restructuring. An economic shift towards these goals—regardless of the scenario—will be a long process and is far from realization (Fennemann et al. 2018). Engagement of applied research can co-determine the direction of CE application possibilities and the discourse on it. This paper aims to discuss the prominent and—until now—disregarded aspects of logistics in CE initialization. The research question this paper seeks to address is thus: Are the current logistics management approaches ready, adjustable or unable to function in a CE? For this purpose, the observed expression of CE is defined, deducing its facets and respective limits to provide scientific precision in the following discussion. The methodological approach of this paper consists of an extensive literature review of the intersection CE and logistics, including reviews of the current and relevant state of research in this field. The selection of the discussed sources was determined by their perceived impact on this dynamic research area, resulting in a method of discussing currently influential research results. Based on these parameters and resources, the field of logistics in CE will be highlighted, analyzing a sample of the currently dominant approaches, models and concepts in logistics regarding their compatibility with CE. The paper concludes with a critical evaluation of the discussed concepts and proposes suggestions for further research.

2 Definition of Circular Economy

To examine the validity of existing logistical practices in the context of a circular economy, the core concepts of involved areas are to be defined. In essence, the definition of a circular economy, which is thereafter used, is a necessary limitation. A multitude of possible outlines for what constitutes a CE exist, often incompatible with each other, as an abstract concept is interpreted differently by various stakeholders. (Corvellec et al. 2022) point out this large variety in their review of related literature. This section aims to define a coherent CE understanding to be applied in this work.

CE in all accepted definitions describes economic activity that abides to principles of efficient resource use. These should guarantee a constant or longest possible use of invested resources and labor through continuous reintroduction of used resources. A collection of up to eleven interconnected principles is often given as the according basis of CE: *Reduce, Re-Use, Repair, Redesign, Remanufacture, Repurpose, Refurbish, Recover, Recycle, Refuse, Rethink* (Lehmacher and Bödecker 2023). The returning prefix "re-", meaning "back" or "again", indicates the circular fashion in which economic activity is to be conducted. Resources, which were extracted and used in production, will not be considered waste and a liability after use, but a secondary resource—ready to be reclaimed via the "Rs". While all aspects of current economies would change in CE, it primarily alters the production and design of products and materials. By keeping the "loop" (in which materials or products are reintroduced into the process) as small as possible, high levels of value in the form of labor and previous processing can be preserved (Vrzel 2022). This leads to a modification of nomenclature, labeling the preliminarily last step of the production process in CE not shipment to a "consumer", but to a temporary "user" of the product, its material and, ultimately, the resource. Figure 1 displays this circular approach while highlighting informational and logistical comprehensiveness.

A further distinction is possible between the circulation of non-organic materials, as used in e.g. technology production, and the circulation of biological resources, which can be considered renewable (Fennemann et al. 2018). While the latter by its nature can comparatively easily be *reused* and *recovered*—for example with constantly recycled water (Lehmacher and Bödecker 2023)—its circularity on a high level of processing is not common, often not possible (Fennemann et al. 2018). Although these renewable, biological materials are considered to be consumed—not temporarily used—the resulting waste can and is partially separated and reintroduced e.g. as fertilizer. The non-organic materials on the other hand rarely circulate at all—hence the introductory disparity of resource scarcity and waste abundance. From

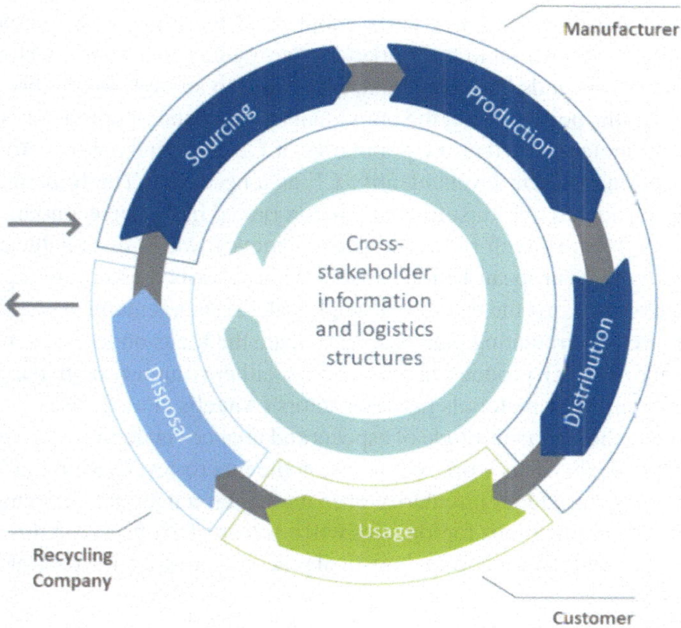

Fig. 1 Visualization of value addition in a CE, based on (reproduced from Fennemann et al. 2018, own translation)

an economic point of view, the use of this waste as a resource is generally far less interesting than using primary resources, as long as these are obtainable and more affordable (Corvellec et al. 2022). This obvious hurdle to circulation of materials can be considered a great economic challenge to implementation of CE aspects by current markets. To further enable the principles of "11 Rs", material purity is a vital factor of CE. Merely resources that can be properly separated in the recovery (down to the highest possible level of reintroduction) are applicable to be circulated. The designing of products therefore requires modularity while avoiding use and release of toxic materials in production and recovery (Angelis et al. 2018).

Based on these aspects, (Grafström and Aasma 2021) identify three possible scale expressions for implementation of CE: The micro-, meso- and macro-level. The last describes a comprehensive vision, involving constant cycling of all materials and resources in an area, as extensive as possible (national, supranational or global). While this is a long-term vision by many proponents of CE, it is the most abstract and distant option for application, as its complexity hinders implementation. Nevertheless, actors like the Ellen MacArthur Foundation see this as the ultimate aim of CE promotion, as global application of this approach is considered to largely stop release of harmful emissions and solve issues regarding waste and resource scarcity (Ellen MacArthur Foundation 2023). The other levels are closer to realization: Single companies (micro-level) or partners in a supply chain (meso-level), which abide by the CE principles as close as economically possible. While this arguably is not solving the global issues at hand, early adopters of these principles could benefit from an established position in the market for "secondary materials", when primary materials run dry. Brands like Patagonia, Inc. claim to strive towards this goal (Oh et al. 2021). At this point in time, the individual and voluntary approach to CE could actively be promoted to further the popularity of CE as defined above. To this end, policy makers on different levels identify CE as a necessity: The limited nature of materials is considered to be insufficiently incorporated in current economic activity. Circular principles are therefore encouraged as a move towards sustainable economic activity, e.g. by the European Union. Their role is described as promoting or sanctioning behavior towards the desired principles of CE (Fennemann et al. 2018). The example of already circulating pallets, to anticipate the focus on logistics, highlights the important role which product standardization, either voluntary or given, possesses for circulating modular materials between actors (Angelis et al. 2018).

To conclude, there is a multitude of aspects and nuances in defining CE, variations of the concept coexist. This work will focus on the macro-level vision of CE, which comprehensively introduces circularity of materials on a national, supranational or global level. The implications for logistics will be clearest to observe in this scenario, including threats and risks if current logistical practices are not adjusted towards their new tasks in CE.

3 Logistics in Circular Economies-State of the Art and Foresight

To date, little research on the role of logistics in CE exists. Two of the most impactful and extensive literature reviews on CE (Merli et al. 2018, Homrich et al. 2018) do not list a single publication that focuses on logistics. A third review paper (Lüdecke-Freund et al. 2019) does mention logistics and predicts that the implementation of circular economies will open up new business opportunities for logistics companies specialized in reverse logistics, thus narrowing down the role of logistics to reverse logistics operations. Finally, (Farooque et al. 2019) does dedicate a whole section to the role of logistics in CE. It concludes that barely any literature on the role of logistics in CE exists to date, beyond a few publications on reverse logistics. This is supported by a comprehensive analysis by (Kirchherr et al. 2023), which finds that "significant reconfiguration of supply chains" (ibid. p. 9) is only mentioned in ten percent of CE definitions. On the other hand, some publications do explicitly focus on logistics operations in—future—CEs. None of these studies suggest that emission from the logistics sector will be reduced in a meaningful way. (Genovese et al. 2017), e.g., which looks at two case studies of CE from the chemical and food industry, concludes that the emissions from the logistics sector itself remained unchanged compared to a linear supply chain. Even worse, (Nasir et al. 2017) and (Dente and Tavasszy 2018), both of which quantitatively compare the emissions of CE with linear supply chains, suggest that the emissions of the logistics sector even increase in a CE. Beyond emissions, the findings in existing studies are similarly discouraging. While (Lopes de Sousa Jabbour et al. 2018; Geissdorfer et al. 2018; Yadav et al. 2020) demand the emergence of new logistics concepts to be fit for CE, they do not propose any such concepts themselves. Further, (Van Buren et al. 2016) concludes that the current legal frameworks of most European economies are actively preventing CE to evolve, e.g. as the processing of goods currently defined as "waste" is legally not permitted in many cases. Yet, logistics will play a pivotal and vital role in CE. Even in the simplest transitions from supply chains towards CE, in which the sink of the previously existing supply chain would link up with the source of the previous supply chain, at least one additional logistical relation—between the sink and the source—is necessary. Even in this very simplified model, one additional logistical relation thus has to be served, increasing the logistical workload in the overall system. In more complex and realistic CE approaches, the increase of demand for logistical services will typically be higher than that. At the same time, barely any concepts to make logistics itself fit into the CE principles seem to exist.

Additional to the lacking presence of logistics in current CE research, the general readiness of logistical means to the ends of CE require discussion. Even though application of CE is not to be expected any time soon, earlier insights on e.g. policy suggest the need for a shift towards preparedness. There are four main means of logistics currently: cargo ships, trucks, trains and cargo planes. None of these are currently operating emission-free and thus would not fulfill the criteria, which CE sets towards any form of transport. Emission-free means of transportation, e.g. hand-drawn carts, cargo bikes or fully electrified vehicles are usually limited in capacity and thus only used for first and last mile logistics, if at all. The outlook for cargo-ships is sinister, with experts not expecting emission-free vessels to penetrate the market before 2035 (De Beukelaer 2021). For long-distance maritime vessels, alternative and emission-free fuels will be required, which are not fully developed, yet (Qazi et al. 2023). Like maritime shipping, aviation has to heavily rely on innovation towards more efficient alternative fuels in future, as fully electric solutions are not available and unlikely to become widely available. Hydrogen solutions are unlikely to be readily available long before 2050, too (Davydenko and Hilbers 2024; Baxter 2021). For trains, (Logan et al. 2021) looked into the current pathways to becoming emission-free in Japan and the UK, two countries strongly committed to emission-free transports by 2050. They conclude that significant investments and policy changes will be required to meet the 2050 target, especially in the UK, which currently still plans to acquire long-lasting diesel engine locomotives until 2040. In both countries, they do not expect to reach emission-freedom before 2050. Finally, on the road, electrified solutions are available already, at least for all but heavy cargo. However, they are only slowly penetrating the market. Even by 2050, when Europe aims to be entirely climate-neutral, some suggestions expect only 40 percent of road freight to be transported emission-free (Ruhnau et al. 2019). Innovations which the sector put a lot of hope in, such as truck automation and platooning, are slowly proving to barely be able to reduce emissions (Paddeu and Denby 2022). Apart from these technological limitations, knowledge about the potential efficiency of climate-reduction strategies is scarce. (Kaack et al. 2018), e.g., studied the potential of modal shift policies and strategies and concluded that "the global potential for reducing greenhouse gas emissions through modal shift remains unknown" (ibid. p. 22).

Given the questionable ability of the transportation sector to become emission-free until 2050, the increasing scarcity and competition of electricity, raw materials needed for batteries and alternative fuels, (De Blas et al. 2020) conclude "that the only way to achieve decarbonization is a profound change in the dominant economic paradigm" (ibid. p. 14). The CE-readiness of logistical means can be considered lacking for the foreseeable future.

4 Compatibility of Three Selected Logistics Concepts with Circular Economy

Until now, logistics is a product of linear economics. It provides goods as requested and is therefore following parameters, downstream from other economic activity. Nevertheless, policy changes towards circular principles e.g. by the EU could be implemented and requested rather abruptly, given the accelerating urgency of sustainable civilizational adjustments. All economic theories, including those in logistics and SCM in general, should in time take inventory of their models' compatibility with CE. The collection below is no complete list of concepts but a selection of influential practices that would be affected by a shift towards CE. Its spotlight analysis in this limited scope stands exemplary for the sum of modern logistics management aspects and their compatibility with CE.

SCOR Model The visualization of complex modern supply chains requires clearly arranged models. A popular depiction is the strategic Supply Chain Operations Reference (SCOR) model, first introduced by the Supply Chain Council (SCC) in 1996. This tool was originally able to display a whole supply network, dividing both internal and external processes of the chain into the sub-processes *source*, *make*, *deliver* and lastly *plan*, which overarches the other three (Huan et al. 2004). The linear chain in this model is visualized from left to right, following the product's route through production. Indicators for returning aspects (from right to left) usually display the desired flow of information or monetary compensation over organizational boundaries. In updated versions of this model, the reintroduction of resources into the process is represented by the additional process *return*. But rather than providing the model language for circular resource allocation, this addition was described by the SCC as an extension towards "post-delivery customer support"—meaning the return of defective or unwanted goods, regardless of the resource's reintroduction into the chain (Supply Chain Council 2012, p. 2). This option represents a currently small amount of backflow, as the short lives of products typically still end with their delivery to a consumer (Fennemann et al. 2018). An option for circularity is not available in the clear-cut model language of this version of SCOR 11. A recently published "Digital Standard" version of the SCOR model puts more emphasis on circular possibilities, looping its processes with each other. Since the return of material is more prominent in the SCOR-DS version, it can be considered a move towards a circular SCOR model (Association for Supply Chain Management 2022). This update again insufficiently includes circularity of secondary materials (it aims to display digital omnichannel supply networks) but improves the possibilities for organizations to do so. Despite the recent update, SCOR 11 remains the most widely used version and thus can still be considered the industry standard for now.

In general, supply chain models display the transformation of an initial resource to the consumer of the final product, via multiple levels of suppliers and a variety of customers—they are often still limited to this one direction. Both the initial source and the final sink of this flowchart will have to be redefined by CE, where sourcing is mainly from used, secondary resources and consuming will give way to temporary use of products and therefore, resources. The SCOR model, while gradually being updated towards such possibilities, could benefit from further exploring and distributing new options to display circular supply chains in the future.

Incoterms Norm In transport logistics, the transition of goods and the associated risk responsibility is a vital element. "International commercial terms" (Incoterms) were formulated to ensure shared definitions and standards in these transitions. First formulated in 1936, these terms influenced especially international trade with a set number of transfer options like Ex Works (EXW) or Free on Board (FOB), the naming of which describes the transfer of ownership and associated legal implications (Durdağ and Delipınar 2020). In CE, the importance of mutually agreed upon delivery standards will not diminish, but rise: Circular coordination in supply chains will include a multitude of pitfalls regarding risk ownership and responsibilities. Existing Product-as-a-Service (PaaS) solutions and their related logistics requirements blur the legal lines between all stakeholders already. Nevertheless, the current options for transfer of ownership given in the latest Incoterm version of 2020 can be considered insufficient for CE purposes: Similar to the long valid SCOR models, the Incoterms assume a linear supply chain, starting "Ex Works" and ending at the latest with unloading at the final destination ("Delivered at Place Unloaded"). They manage the final handover of risk in one direction—there are no explicit Incoterms options for return logistics, much less circular movements of goods and risks, from purposes of repair or reintroduction of resources. As ownership of products for many goods will not transfer completely in CE (PaaS), besides business models (Morseletto 2020), related Incoterms require revision and development, too. This task is impossible to be completed before policy on CE is in effect, as legality sets the stage for relationships in supply chains.

In general, new standards will be necessary in CE (Fennemann et al. 2018), and logistics standards are not excluded from this change. A legal framework cannot yet be formulated, as parameters of CE implementation are not given (and will not be, anytime soon). A constant calibration of current agreed upon norms for changes in resource circularity is nevertheless necessary, as the stakeholders of the supply chain require sound and reliable options for the transportation of goods.

Lean Logistics There are currently established trends that seem close to the principles of the CE. As its goal is to avoid all waste, it should have similarities to the concept of lean logistics management. Emerging from manufacturing in Japan's automotive industry, lean manufacturing described techniques of efficient resource use to avoid waste ("Muda") and provide products at lowest possible costs through high productivity (Kumar et al. 2022). The economic success of this trend and its international spread towards other areas could indicate the successful merger and arrival of CE-ready economic activity, without explicit policy or guideline. In logistics, this leads to derived principles like synchronization of material flows, pull principle or standardization of flows to avoid disturbances (Klug 2018). The included coordination across the supply chain of lean management is compatible with the CE's need for transparent communication between actors. Nevertheless, lean logistics is not synonymous with CE-ready logistics: The interpretation of waste, which is minimized in lean logistics, is manifold. Material waste is merely considered one of multiple forms of Muda (Klug 2018; Salah et al. 2022). Accordingly, waste in lean management is not the production of material waste but the unnecessary use of monetary equivalents. The indifference of the lean concept to material waste in general becomes evident in its missing integration of returning resources or product designs based on principles of material reintegration. As lean production, management and logistics intend to maximize the output and profit of a single business or supply chain, its waste avoidance is not aimed at the same goal as CE; the attention span of which doesn't end with the sale.

Current trends in logistics are not necessarily opposed to CE. On the contrary, sustainable organization and CE as its expression can be considered trends as well. While the societal concern of unsustainable use of resources manifests in trends to boost efficiency, the profit motive (or necessity) to date remains the most prioritized factor for management, production and logistics. Therefore, the lean trend in these fields cannot be considered a move towards CE—at the same time, it is not incompatible either. An implementation of CE on micro-, meso- or macro-level would for certain include lean principles: the earlier CE example of Patagonia, Inc. recommends lean management principles as well (Patagonia, Inc. 2012). The profit motive will for the foreseeable future be the leading incentive for private organizations to implement CE or at least some aspect of CE principles. Its marriage with CE can be considered a major task for policy makers.

The discussed concepts are not yet fully compatible with CE, as logistics is built around linear economies: It follows the current de-facto norm of a one-way, start-to-end process. Circular motions in logistics are outliers and do not dictate according practices. The need for complex logistics models that can display and practically support CE, calls for advanced technology innovation, e.g. in tracking and tracing. Likewise, the increased importance of fostering inter-organizational communication across the (circular) supply chain can be deduced from the analysis. CE compatible models therefore require both: improvements in technology and collaborative adjustments. Corresponding motivations will be the base for logistical solutions, ready for circular supply chains. As (Fennemann et al. 2018) put it: "The potential of CE can only be realized with *fitting* logistical solutions" (ibid. p. 15, own translation and emphasis).

5 Conclusion

This paper examined the role of logistics in CE and briefly analyzed the compatibility of current practices in logistics with the concept of CE. The scope of this paper allowed for a sample of supply chain practices. The authors expect similar shortcomings for other practices, but had to limit this discussion paper to those discussed above. Further, the framework limits the possible research options in mitigating the identified gap of current SCM and CE. Associated possibilities for further research are discussed below in more detail.

Current CE research focuses on the production aspects of circularity, logistics is considered an afterthought, its adaptation is assumed. Logistical means are revised and increasingly improved, to taper harmful distribution methods revolving around the combustion engine. Nevertheless, logistical infrastructure is far from being able to provide emission-free transportation, which would be required in CE. The logistical ends of providing CE-ready solutions are insufficient as well: Circular supply chains, PaaS and other implications of CE will require new logistics practices. This is reflected by the compatibility analysis of existing concepts. The three discussed current influential concepts provided a clear indication that linear logistics (and more comprehensive, SCM) will not be able to display the realities of CE, as is. There will be no national or supranational application of CE in the near future, the current incompatibility of linear concepts with circular visions is not surprising—nevertheless, research into the capability of logistics networks to satisfy its needs has to be started before its application is necessary. Logistics practitioners will indeed be on the receiving end of CE application: They cater to demand of the economy, which can change quickly. They are in many ways equipped to distribute according to CE principles, little general overhaul of infrastructure (apart from its move towards sustainable transportation) is needed. However, current approaches in SCM and therefore logistics are a product of linear economics and are formulated accordingly. Its models, strategies and tools require revision for a coming circularity of economic activity. For this purpose, it is insufficient to form circles from the current chains by fusing their ends to the start, as a multitude of parameters is subject to change in the expected transformation. Coordination and the flow of information along the supply chain will increase sufficiently through circulation of materials, further complicating already complex supply networks. The requirements CE has for products are mainly circulating around necessary changes in design. With these adjustments towards e.g. modularity, smaller circles in the re-use of a product lead to a conservation of its resources and the labor that went into its manufacturing. Similarly, the design of supply chains and its models and tools require a redesign to be compatible with a circular approach; conventional linear designs were demonstrated to be lacking. Comprehensive proposals for more circular and therefore more sustainable management of economic aspects have to take this necessity into account.

Macro-level CE is still an abstract idea with many challenging hurdles and outstanding definitions to be established and agreed upon. Based on its circular models in nature, e.g. the water cycle, its advent is sometimes expected to occur effortlessly, as a natural evolutionary step of an economy in crisis. After analyzing the current trends in CE applications, this bold claim has to be challenged: CE and its moving parts, like SCM models, are to be carefully and precisely designed by trendsetting forces, i.e., policy of supranational entities. As long as a linear economy is more economic to the directly influencing stakeholders, intervention policies will be needed to establish CE principles. The single use of resources and production of waste is the norm—recycling, waste reintegration et al., including their logistical backdrop, are still considered outliers. Comprehensive economic restructuring in steps is needed to proximate CE. Too slow, and spaceship earth fails its mission of sustainable maintenance.

CE is mostly a vision for comprehensive changes, necessary to maintain economic activity in the long-run. It requires a level of global cooperation and coordination that is currently not given and implies vast changes to business models and supply chains (Morseletto 2020; Angelis et al. 2018). Potentially, the conflict of objectives between responsible use of resources and current trends in economies (e.g. fast fashion, single use electronic items) cannot be solved by market actors alone or be left to the consumer, who often is intentionally deceived by green washing or economically inclined (or pressured) to go for the most affordable option. Meaningful policy frameworks, while restricting private businesses' range of motion, can lead the way towards politically intended behavior, e.g. in bodies like the European Union. Accordingly, policy changes are needed (not merely regarding logistics), guiding all markets towards the desired practices. For this purpose, a shared definition of CE—at least on EU level—is necessary, including targets, responsibilities and deadlines. Further, purposive public procurement by the EU or national governments can be of practical relevance in promoting CE principles (Angelis et al. 2018; Fennemann et al. 2018). As (Angelis et al. 2018) conclude, there is little research results yet on how the practical implementation of CE should be conducted. Nevertheless, the current trajectory for fast atmospheric heating makes sustainability policy a global necessity, not a "nice-to-have" initiative by a political faction. To realistically move towards CE, its compatibility with current economic models has to be researched comprehensively, as e.g. (Corvellec et al. 2022) see CE under current circumstances to be little more than window dressing: Efforts by the EU to create win-win situations rather than introduce restrictions or trade-offs for private businesses are deemed insufficient and too vague for practical implementation—as an umbrella concept, it causes excitement without a clear path. This is evident by foundations that collect data and spread knowledge on the topic of CE, but usually stop short of suggesting policy that would imply prohibition or taxing towards circular targets. Their financing by private companies indicates short-term private interests, which could interfere with extensive and potentially costly changes towards CE. Rare occasions of companies, which move their production and supply chains towards circular efforts (despite higher short-term expenses) can showcase the individual benefits of being "CE-ready" (Lehmacher and Bödecker 2023). In the long run, the maintenance of past investments in labor and resource processing through circulation of materials could be profitable—but since profits can be expected to be slim in the advent of comprehensive CE application, a legal framework is needed to initiate obligatory aspects of circularity. (Fennemann et al. 2018) suggest a taxation of resource use as a policy direction, to encourage careful allocation.

Future research has to tackle the discussed incompatibilities of circular economies and linear logistics. Currently, common practices and tools are unable to sufficiently display and therefore manage CE supply chains. This work recommends further engagement with circular material utilization in sourcing, production and repeated distribution, to preserve earth's valuable resources.

References

Association for Supply Chain Management (2022) Supply chain operations reference model-SCOR digital standard. https://www.ascm.org/globalassets/ascm_website_assets/docs/intro-and-front-matter-scor-digital-standard2.pdf. Accessed 22 May 2024

Angelis R, Howard M, Miemczyk J (2018) Supply chain management and the circular economy: towards the circular supply chain. Prod Plan Control 29(6):425–437. https://doi.org/10.1080/09537287.2018.1449244

Baxter GP (2021) Decarbonizing International air cargo transportation's carbon footprint: A review of the world air cargo carrying airlines current and potential environment related measures and strategies. Int J Environ Agric Biotechnol 6(6):265–290. https://doi.org/10.22161/ijeab.66.31

Corvellec H, Stowell A, Johansson N (2022) Critiques of the circular economy. J Ind Ecol 26(2):421–432. https://doi.org/10.1111/jiec.13187

Davydenko I, Hilbers H (2024) Decarbonization paths for the Dutch aviation sector. Sustainability 16(3):950–964. https://doi.org/10.3390/su16030950

De Beukelaer C (2021) Tack to the future: Is wind propulsion an ecomodernist or degrowth way to decarbonise maritime cargo transport? Clim Policy 22(3):310–319. https://doi.org/10.1080/14693062.2021.1989362

De Blas I, Mediavilla M, Capellán-Pérez I, Duce C (2020) The limits of transport decarbonization under the current growth paradigm. Energ Strat Rev 32:100543. https://doi.org/10.1016/j.esr.2020.100543

Dente SM, Tavasszy LA (2018) Impacts of trade related sustainability strategies on freight transportation: Modelling framework and application for France. Transp Res Part D: Transp Environ 58:308–319. https://doi.org/10.1016/j.trd.2017.04.006

Durdağ C, Delipınar GE (2020) The past, today and future of incoterms in international delivery: A review on the innovations in logistics. J Econ Libr 7(4):201–207

Ellen MacArthur Foundation (2023) Building a circular supply chain. https://www.ellenmacarthurfoundation.org/circular-supply-chains. Accessed 22 May 2024

Farooque M, Zhang A, Thürer M, Qu T, Huisingh D (2019) Circular supply chain management: A definition and structured literature review. J Clean Prod 228:882–900. https://doi.org/10.1016/j.jclepro.2019.04.303

Fennemann V, Hohaus C, Kopka J (2018) Circular Economy Logistics: Für eine Kreislaufwirtschaft 4.0, Fraunhofer IML. https://doi.org/10.24406/IML-N-491576

Geissdoerfer M, Morioka SN, de Carvalho MM, Evans S (2018) Business models and supply chains for the circular economy. J Clean Prod 190:712–721. https://doi.org/10.1016/j.jclepro.2018.04.159

Genovese A, Acquaye AA, Figueroa A, Koh SL (2017) Sustainable supply chain management and the transition towards a circular economy: Evidence and some applications. Omega 66:344–357. https://doi.org/10.1016/j.omega.2015.05.015

Grafström J, Aasma S (2021) Breaking circular economy barriers. J Clean Prod 292:126002. https://doi.org/10.1016/j.jclepro.2021.126002

Haas W, Krausmann F, Wiedenhofer D, Lauk C, Mayer A (2020) Spaceship earth's odyssey to a circular economy-a century long perspective. Resour Conserv Recycl 163:105076. https://doi.org/10.1016/j.resconrec.2020.105076

Homrich AS, Galvão G, Abadia LG, Carvalho MM (2018) The circular economy umbrella: Trends and gaps on integrating pathways. J Clean Prod 175:525–543. https://doi.org/10.1016/j.jclepro.2017.11.064

Huan SH, Sheoran SK, Wang G (2004) A review and analysis of supply chain operations reference (SCOR) model. Supply Chain Manag 9(1):23–29. https://doi.org/10.1108/13598540410517557

Kaack LH, Vaishnav P, Morgan MG, Azevedo IL, Rai S (2018) Decarbonizing intraregional freight systems with a focus on modal shirt. Environ Res Lett 13:083001. https://doi.org/10.1088/1748-9326/aad56c

Kirchherr J, Yang NN, Schulze-Spüntrup F, Heerink MJ, Hartley K (2023) Conceptualizing the circular economy (Revisited): An Analysis of 221 definitions. Resour Conserv Recycl 194:107001. https://doi.org/10.1016/j.resconrec.2023.107001

Klug F (2018) Lean Logistics. In: Klug F (ed) Logistikmanagement in der Automobilindustrie, 2nd edn. Springer Vieweg, Berlin, Heidelberg, pp 287–326. https://doi.org/10.1007/978-3-662-55873-7_7

Kumar N, Shahzeb Hasan S, Srivastava K, Akhtar R, Kumar Yadav R, Choubey V (2022) Lean manufacturing techniques and its implementation: A review. Mater Today: Proc 64:1188–1192. https://doi.org/10.1016/j.matpr.2022.03.481

Lamboll RD, Nicholls ZRJ, Smith CJ, Kikstra JS, Byers E, Rogeli J (2023) Assessing the size and uncertainty of remaining carbon budgets. Nat Clim Chang 13:1360–1367. https://doi.org/10.1038/s41558-023-01848-5

Lehmacher W, Bödecker J (2023) Circular Economy-7. Industrielle Revolution: Der Weg zu mehr Nachhaltigkeit durch Kreislaufwirtschaft. Springer Gabler, Wiesbaden

Logan KG, Nelson JD, McLellan BC, Hastings A (2021) Japan and the UK: Emission predictions of electric and hydrogen trains to 2050. Transp Res Interdiscip Perspect 10:100344. https://doi.org/10.1016/j.trip.2021.100344

Lopes de Sousa Jabbour AB, Jabbour CJC, Godinho Filho M, Roubaud D (2018) Industry 4.0 and the circular economy: A proposed research agenda and original roadmap for sustainable operations. Ann Oper Res 270:273–286. https://doi.org/10.1007/s10479-018-2772-8

Lüdeke-Freund F, Gold S, Bocken NM (2019) A review and typology of circular economy business model patterns. J Ind Ecol 23(1):36–61. https://doi.org/10.1111/jiec.12763

Mendelson R, Dinar A, Williams L (2006) The distributional impact of climate change on rich and poor countries. Environ Dev Econ 11(2):159–178. https://doi.org/10.1017/S1355770X05002755

Merli R, Preziosi M, Acampora A (2018) How do scholars approach the circular economy? A systematic literature review. J Clean Prod 178:703–722. https://doi.org/10.1016/j.jclepro.2017.12.112

Morseletto P (2020) Targets for a circular economy. Resour Conserv Recycl 153:1–12. https://doi.org/10.1016/j.resconrec.2019.104553

Nasir MHA, Genovese A, Acquaye AA, Koh SCL, Yamoah F (2017) Comparing linear and circular supply chains: A case study from the construction industry. Int J Prod Econ 183:443–457. https://doi.org/10.1016/j.ijpe.2016.06.008

Oh W, Chang Y, Park J, Han S (2021) Toward a sustainable paradigm: circular economy solutions in the fashion industry. In: Jakobsen S, Lauvås T, Quatraro F, Rasmussen E, Steinmo M (ed) Research Handbook of Innovation for a Circular Economy. Edward Elgar Publishing, Cheltenham, pp 47–58. https://doi.org/10.4337/9781800373099.00012

Paddeu D, Denby J (2022) Decarbonising road freight: Is truck automation and platooning an opportunity? Clean Technol Environ Policy 24(4):1021–1035. https://doi.org/10.1007/s10098-020-02020-9

Patagonia Inc (2012) Green Business practices. https://www.patagonia.com/on/demandware.static/Sites-patagonia-us-Site/Library-Sites-PatagoniaShared/en_US/PDF-US/green_and_business.pdf. Accessed 22 May 2024

Qazi S, Venugopal P, Rietveld G, Batista Soeiro T, Shipurkar U, Grasman A, Watson AJ, Wheeler P (2023) Powering Maritime: Challenges and prospects in ship electrification. IEEE Electrification Mag 11(2):74–87. https://doi.org/10.1109/MELE.2023.3264926

Ritchie H (2020) Cars, planes, trains: where do CO_2 emissions from transport come from? https://ourworldindata.org/co2-emissions-from-transport. Accessed 22 May 2024

Ruhnau O, Bannik S, Otten S, Praktiknjo A, Robinius M (2019) Direct or indirect electrification? A review of heat generation and road transport decarbonisation scenarios for Germany 2050. Energy 166:989–999. https://doi.org/10.1016/j.energy.2018.10.114

Salah O, Ikram AH, Ali H (2022) The impact of lean and green supply chain practices on sustainability: literature review and conceptual framework. LogForum 18(1):1–13. https://doi.org/10.17270/J.LOG.2022.684

Sosnowski P (2022). Green concepts in the supply chain. Logforum 18(1):2. https://doi.org/10.17270/J.LOG.2022.680

Supply Chain Council (2012) Supply Chain Operations Reference Model-Revision 11.0. http://economia.uniroma2.it/public/ba/files/SCOR11PDF_(1).pdf. Accessed 22 May 2024

Van Buren N, Demmers M, Van der Heijden R, Witlox F (2016) Towards a circular economy: The role of Dutch logistics industries and governments. Sustainability 8(7):647–664. https://doi.org/10.3390/su8070647

Vrzel J (2022) The circular economy basics series-The Technical Cycle. https://www.circularinnovationlab.com/post/the-circular-economy-basics-series-the-technical-cycle. Accessed 5 June 2024

Welsby D, Price J, Pye S, Ekins P (2021) Unextractable fossil fuels in a 1.5° C world. Nature 597:230–234. https://doi.org/10.1038/s41586-021-03821-8

Yadav G, Luthra S, Jakhar SK, Mangla SK, Rai DP (2020) A framework to overcome sustainable supply chain challenges through solution measures of industry 4.0 and circular economy: An automotive case. J Clean Prod 254:120112. https://doi.org/10.1016/j.jclepro.2020.120112

Greening of e-Commerce Logistics by Digitalizing

Arkadiusz Kawa

Abstract The dynamic growth of e-commerce and digital transformation are reshaping logistics, while simultaneously presenting significant environmental challenges. E-commerce logistics is responsible for substantial greenhouse gas emissions, air pollution in cities, excessive energy consumption in warehouses, and the generation of hard-to-recycle packaging waste. This process is further exacerbated by impulsive online shopping and high return rates. Digitalization, combined with the concept of green logistics and the circular economy, offers solutions for sustainable development. Technologies such as IoT, AI, and blockchain support more efficient resource management, optimization of delivery routes, and reduction of CO_2 emissions. They help to reduce the negative impact on the environment. Despite the potential of digital technologies, their implementation requires careful analysis of costs and limitations. A key challenge remains the integration of these solutions in practice, while considering the growing customer expectations regarding environmental responsibility. The digital transformation of logistics is a crucial step toward achieving global sustainability and carbon neutrality goals. This chapter analyzes the role of digitalization in building sustainable development in the e-commerce sector, highlighting both the potential opportunities and risks associated with the implementation of new technologies. To this end, a literature review method was used.

Keywords e-Commerce · Logistics · Sustainability · Digitalization

A. Kawa (✉)
Poznań School of Logistics, Poznań, Poland
e-mail: arkadiusz.kawa@wsl.com.pl

Poznań University of Economics and Business, Poznań, Poland

A. Kolinski and M. Adamczak (eds.), *Digitalisation of the Greening Supply Chain*, EcoProduction, https://doi.org/10.1007/978-3-031-88918-9_10

1 Introduction

E-commerce is one of the fastest-growing sectors in the modern economy (Munoz et al. 2023). With its accessibility, flexibility, and global reach, it has revolutionized the way people buy and sell products. For customers, it offers convenience-the ability to order goods from anywhere at any time, often with access to a wider range of products than traditional stores provide. For sellers, it provides an opportunity to reach a larger audience while reducing costs associated with maintaining physical storefronts. However, the growth of e-commerce comes with challenges, one of the most significant being logistics (Taniguchi et al. 2016).

Products ordered online must be delivered to customers, requiring a well-organized logistics process. In traditional retail, customers visit a store and select products themselves. In e-commerce, the process is far more complex-the product must be received into a warehouse, stored, assembled, packaged, and shipped (Kawa and Pierański 2021). Additionally, many companies must manage returns, which in many countries are legally regulated and often driven by customers' high expectations for shopping flexibility (Shi et al. 2020).

In recent years, a critical aspect of e-commerce development has been its environmental impact. Logistics, in particular, directly affects the environment through energy consumption, carbon dioxide emissions, and waste production. Last-mile deliveries (the final stage of getting products to customers) generate high costs and significant CO_2 emissions, especially when poorly organized. Innovations such as electric delivery vehicles, drones, and advanced route optimization algorithms can provide solutions by streamlining delivery processes (Mangiaracina et al. 2015).

Increasing environmental awareness among consumers further influences companies' actions. Customers are paying more attention to the origin of products, the materials they are made of, and how they are produced and transported. In response, companies are making changes, both independently and through platforms like re-commerce, which facilitate the sale of second-hand products.

Digitalization plays a crucial role in addressing the logistics and environmental challenges of the e-commerce sector. It eliminates paper documentation, enables precise process monitoring, and allows for efficient optimization. By leveraging real-time data exchange, inventory visibility, and transportation optimization, companies can reduce costs and their environmental footprint. Data-driven management systems, such as just-in-time approaches, help avoid overproduction, reduce the need for excess inventory, and decrease energy consumption. Digital tools are also employed in managing returns, facilitating product recovery and reintegration into the supply chain (Adeniran et al. 2024).

However, digitalization is not without its challenges. While energy management systems based on advanced technologies can reduce operational energy use, the energy demand for servers, data centers and advanced software continues to grow. Digital transformation is therefore both an opportunity for increased efficiency and sustainable growth, and a challenge that requires a responsible approach to technology implementation.

This chapter aims to highlight the role of digitalization in the sustainable development of e-commerce and to present its impact-both positive and negative-on the sector. For this purpose, a method based on a review of the literature was applied.

The structure of the chapter is as follows. Section 2 presents a literature review focusing on digitalization, sustainable development, and the impact of technologies on e-commerce logistics. An analysis of various aspects of the environmental impact of e-commerce logistics, such as greenhouse gas emissions, urban air pollution, energy consumption in warehouses, and the generation of packaging waste is given in Sect. 3. Section 4 provides a discussion of the concept of green logistics and the role of digitalization in improving efficiency and sustainability in logistics. After that various technologies and concepts, such as Industry 4.0, the Internet of Things (IoT), paperless solutions, and blockchain, and their application in e-commerce logistics are described. Section 6 discusses the paradox of digitalization's impact on sustainable e-commerce logistics. The role of regulations in promoting and enforcing environmentally and socially responsible practices in e-commerce logistics is delivered in Sect. 7. Finally, a summary of the main findings from the research is presented.

2 State of the Art

In recent years, digital transformation has become a key element in the development of e-commerce and logistics, which has generated widespread interest among researchers. Digitalization is changing traditional business models, enabling faster and easier data acquisition, processing and analysis. At the same time, questions are being raised about the impact of this process, not only in economic terms, but also in environmental terms.

There is a growing body of literature that brings together the topic of the role of information and communication technology (ICT) and digitalization and sustainability, but without the context of e-commerce (Lange and Santarius 2020). On the other hand, the literature on sustainability and e-commerce is rich and growing, but few publications focus on the role of digitization and its impact on sustainable e-commerce. With the rise of digitalization in the e-commerce area comes many challenges related to energy efficiency, CO_2 emissions and resource management. Digital transformation is leading to changes in supply chains, with significant environmental impacts. The first studies are beginning to appear in the literature, in which authors analyze whether digital technologies can support the sustainability of e-commerce and how responsible digitization can contribute to corporate social responsibility (CSR) goals (Cardinali and De Giovanni 2022).

For years, the logistics sector has been facing environmental and economic challenges related to excessive energy consumption and CO_2 emissions (Bishop 2019; Khan et al. 2020). More and more attention is being paid to the concept of green logistics, which is based on the use of modern technologies to optimize energy consumption and reduce environmental impact. Digitalization, coupled with the concept of a circular economy, is an important direction for research on reducing the carbon footprint of the logistics industry, both at national and global level (Kwilinski et al. 2023).

Researchers indicate that ICT-based systems can contribute to sustainable development by reducing waste and using resources more efficiently (Yunlin 2023). One of these is transport, which is one of the main consumers of energy. According to Tsakalidis et al. (2020), digital technologies such as artificial intelligence (AI) and the internet of things (IoT) enable advanced route planning and optimization, significantly reducing CO_2 emissions. Research work in this area points to the potential of digitalization in modernizing transport policy and developing intelligent traffic management systems that support shared transport and electromobility. Digital infrastructure and relevant technologies can also be used to improve energy efficiency, for example through dynamic management of electricity distribution (Tsakalidis et al. 2020; Bishop 2019; Kwilinski et al. 2023).

Digital tools such as IoT, blockchain and quantum technology are being used to support green initiatives in supply chains at different levels: from international operations to closed recycling loops (Nham 2022). The integration of these technologies can provide synergistic benefits and their application in green supply chains is showing promising results, but is still at an early stage of development. According to Mulholland et al. (2015), further exploration of the potential of combining different technologies is needed to achieve greater potential for positive environmental impacts, as well as an understanding of the limitations and costs associated with their implementation (Tsakalidis et al. 2020).

Digitization not only enables better internal management within a company but also facilitates collaboration with other entities or simplifies it. An example of this is the use of ICT tools in the sharing economy. Ortega et al. (2023) demonstrated in their research that the application of digital technologies in car-sharing can reduce CO_2 emissions from urban transportation systems and contribute to more sustainable urban development.

Digital sustainability has a special role in European Union (EU) policy, which has the ambitious goal of becoming the first carbon-neutral region by 2050. In 2019. The EU adopted the Green Deal policy, which aims to promote prosperity by eliminating the negative environmental impacts caused by economic development through the provision of, among other things, innovative green technologies.

To summarize the literature review, digitalization offers extensive opportunities for the sustainable development of the e-commerce logistics sector. While the potential of technologies such as AI, IoT and blockchain in this area is promising, further research is needed to better understand their limitations and costs, as well as to develop integrated, effective solutions in practice.

3 The Impact of e-commerce Logistics on the Environment

The rapid growth of e-commerce brings with it significant environmental challenges (Al Mashalah et al. 2022). The logistics sector that supports it is mainly powered by fossil fuels, which has a major impact on greenhouse gas emissions and exacerbating climate problems. Currently, this sector accounts for around 24% of global CO_2 emissions, and the European Environment Agency forecasts that this share could rise to as much as 40% by 2050 if effective mitigation measures are not implemented (Yunlin 2023).

Kong et al. (2024) identified key areas of environmental impact of e-commerce logistics, which can be supplemented with other aspects:

1. Greenhouse gas emissions and climate change. One of the most significant sources of greenhouse gas emissions in e-commerce logistics is transport, which requires a large amount of fossil fuels, especially for last mile deliveries. Increased delivery vehicle traffic, especially in cities, leads to increased CO_2 and other greenhouse gas emissions, which contribute to global warming and climate change, and reduced quality of life for residents. This is particularly evident during busy online shopping periods, such as holidays or special sales, when the number of deliveries increases dramatically. A challenge is the fragmentation of deliveries; shipments are often small and contain a variety of products, complicating the shipping process. When there are many deliveries, especially in residential areas, the lack of efficient last-mile delivery systems results in significantly longer distances travelled (Jurburg et al. 2023).
2. Urban air pollution. The increasing number of deliveries generates intensive vehicle traffic in urban areas, resulting in air pollution. Exhaust emissions containing hazardous substances have a negative impact on the health of urban residents and reduce air quality. These problems are exacerbated in areas of high traffic density, causing a greater number of respiratory and cardiovascular diseases.
3. Energy consumption in warehouses. Logistics operations related to products in warehouses are energy intensive. In many cases, this energy comes from non-renewable sources, which increases the carbon footprint of the sector. Warehouses require energy both to maintain the right temperature for storing goods and to power automatic sorting systems. As a result, each online order contributes to an increase in energy demand, which is an environmental challenge.
4. Generation of packaging waste. One of the problems of e-commerce logistics is the generation of a large amount of packaging waste. Cardboard packaging, plastic protective films and cushioning materials such as foam or polystyrene become waste once the product is delivered. Often these materials are difficult to recycle or are not recyclable, resulting in them ending up in landfills where they take a long time to decompose. This problem is exacerbated by overpackaging, which increases the volume of waste and puts additional strain on the environment.

Although online shopping has its downsides, it may still have a smaller environmental footprint compared to traditional in-store shopping. Research by OliverWyman (2021) shows that the delivery of products ordered online generates less CO_2 emissions than a consumer going to store. The biggest challenge is the energy consumption of buildings. Offline sales require a lot of energy for lighting, heating, etc. of the display area. According to the study, buildings used for e-commerce produce several times less CO_2 emissions than physical retail. It is therefore not clear which type of shopping has a greater impact on the environment. Much depends on the type of products, how often they are bought, how they are delivered, etc.

Notwithstanding, a very big problem is the impulsiveness of online shopping. The aforementioned convenience of buying at anytime from anywhere means that purchases are often ill-considered. Impulsively purchased products are often of poor quality, which quickly makes them lose value and become waste. Such purchases are associated with an increasing number of shipments-each order means additional transport, which generates CO_2 emissions, especially in the case of express deliveries. They also contribute to the overconsumption of packaging. Returns exacerbate the problem-each returned parcel means another consumption of energy and resources, and some of these products do not go back on sale. Impulsive buying also affects overproduction, as increased demand encourages companies to produce more goods. All of this leads to unnecessary waste of natural resources and additional pollution of the environment, showing that seemingly innocent purchases have serious consequences for our planet.

On the other hand, some consumers are increasingly aware of (Štofejová et al. 2023) and expect sustainability measures from companies, which forces companies to look for new solutions.

4 Green Logistics and its Digitalization

The circular economy is one of the key elements of the contemporary economic transformation, redefining the traditional approach to organizing logistics processes. Within it, reverse logistics, alongside traditional supply chains, plays an increasingly important role. Reverse logistics deals with the processes of managing the flow of goods and materials from the customer back to the manufacturer or other points in the supply chain. It focuses on managing product returns and recycling to create closed lifecycles for goods, reducing waste and increasing resource efficiency.

A complementary concept to reverse logistics is green logistics, which focuses on minimizing the negative environmental impact of logistics processes. It has become a key area of interest for stakeholders seeking to promote sustainability. These stakeholders include logistics professionals, business owners, policy makers, researchers and consumers. Green logistics is based on the use of innovative solutions to reduce carbon emissions and increase process efficiency. It is an approach that combines economic, environmental and social objectives, ensuring their harmonious balance (Yunlin 2023).

The digitalization of logistics is further reinforcing this. It is transforming global supply chains, creating opportunities to increase efficiency, reduce costs and improve sustainability. The use of ICT plays an important role in reducing transaction costs and enabling the digital delivery of some goods and services. For example, automation of customs declarations reduces customs clearance times and transit times for goods. Access to ICT platforms allows sellers, especially from developing countries, to reach more consumers in domestic and foreign markets, acting accurately and often more cheaply than through traditional methods (Sun et al. 2021).

Digitization and the "greening" of the economy are two key drivers for high-quality growth and sustainable economic development. Research by Xie and Wang (2023) in China shows a complex relationship between the digitization of logistics and carbon emissions. The impact of digitalization on carbon intensity is non-linear and takes an inverse shape. The level of transformation and modernization of the logistics sector plays a mediating role, positively influencing the reduction of carbon intensity. According to the authors, achieving the "double zero carbon" target is not possible without the support of the digital economy.

The digital transformation of logistics, based on IoT technology and big data analytics, can improve transport and warehousing processes and reduce the amount of packaging materials used. The use of technologies such as geolocalization and big data allows more efficient delivery routes to be planned, reducing fuel consumption and emissions. The introduction of electric vehicles into delivery fleets is also helping to reduce emissions, particularly in urban areas. These solutions are increasingly being used in cities, where the transport load is highest (Kwilinski et al. 2023; Masood and Alam 2020). Furthermore, by analyzing the size of the products, the system allows packaging to be tailored to the individual requirements of each order, reducing redundant materials and lowering transport costs.

Companies that deliver goods ordered online can use a crowdsourcing model in which individuals deliver parcels on their way to their daily destinations. An example of this is Amazon Flex. It is a tool that connects individuals who want to deliver parcels with Amazon. It does not require investment in vehicles, which reduces the burden on infrastructure and increases delivery flexibility. Crowdshipping, which involves the community in making deliveries, allows more efficient use of existing transport resources, which reduces CO_2 emissions (Kwilinski et al. 2023). Courier companies can also share a fleet of vehicles, which reduces the number of unused cars and maintenance costs. For example, one company can share its vehicles with another company during peak hours when delivery demand is highest.

A significant problem in e-commerce is energy-intensive data centers, which are the backbone of online retailers' operations. Heating, ventilation, and air conditioning (HVAC) system in such centers can consume as much as 35–40% of total energy consumption, requiring the use of modern technologies such as high-density cooling units capable of reducing cooling costs by up to 30%. In addition, the implementation of software-defined infrastructure enables better management of IT resources, reduction of energy waste and optimization of processes. Other innovations, such as the use of thermal waste and on-site energy production, further support energy efficiency measures (Masood and Alam 2020).

E-commerce in the digital age is facing energy and environmental challenges. Implementing energy-efficient technologies, optimizing logistics processes and adopting sustainable practices is not only a step towards cost reduction, but also a response to growing customer expectations. Companies that prioritize sustainability have the opportunity not only to reduce their environmental footprint, but also to build a sustainable and positive image in a rapidly changing world.

An important element of the sustainability strategy is "green" IT, involving the design of energy-efficient equipment and systems, as well as the responsible disposal of technological waste. Investment in such solutions not only reduces energy consumption, but also reduces the negative impact of operations on the environment (Masood and Alam 2020).

Monitoring carbon emissions is helpful in preventing e-commerce environmental problems. By measuring, it is easier to analyze and therefore better manage these emissions. Nowadays, sophisticated systems allow companies to measure and monitor emissions throughout the supply chain, making it possible to accurately determine environmental impact. Companies seeking to reduce their carbon footprint can use this data to implement offsetting initiatives, such as tree planting or investment in renewable energy, thereby fitting in with global sustainability goals (Masa'deh et al. 2024).

It is important to remember that the implementation of digital technologies requires a comprehensive analysis, covering the complex regulatory, tax, audit, risk and compliance aspects associated with global business, as well as taking into account environmental and social issues (Cardinali and De Giovanni 2022).

5 Examples of Concepts and Technologies Supporting Green e-commerce Logistics

Industry 4.0

Industry 4.0, also referred to as the fourth industrial revolution, is the concept of integrating digital technologies with traditional industrial processes both within and between organizations (Sarkis et al. 2021). Although Industry 4.0 is a relatively new phenomenon in the literature, automation and digitization were already important elements of earlier industrial revolutions. Key elements of this approach include automation, AI, IoT, real-time data analytics and machine-to-machine (M2M) communication. It aims to create smart factories that are more flexible, efficient and adaptable to rapidly changing market needs. In this way, Industry 4.0 supports the transformation of the economy, enabling sustainable, customized production based on data analytics. Blockchain plays a key role, increasing security, traceability and process efficiency in courier supply chains (Masa'deh et al. 2024). Wang et al. (2023) pointed out that Industry 4.0, being a key factor in digital development, combines different technologies, emphasizing ecological and socially responsible transformation.

Internet of things

A solution that is helpful for e-commerce logistics in the context of sustainability is IoT. It is a network of connected devices equipped with sensors that exchange data with each other. By automating processes and providing precise access to data, IoT enables better delivery planning, improved customer service and reduced environmental impact. IoT devices monitor the entire life cycle of products and assess material sources, supporting companies in using authentic renewable raw materials. Sensors record the location, temperature and humidity of packages, minimizing the risk of damage and increasing efficiency. Real-time information facilitates decision-making related to storage conditions and purchasing and procurement, which supports sustainable business practices (Sarkis et al. 2021).

Technologies such as GPS, RFID and blockchain, combined with IoT, are creating an ecosystem to support green supply chains. Data collected by IoT devices enables the design of intelligent transport systems that reduce emissions, improve the accuracy of shipment planning and allow real-time monitoring of products. Sensors support recycling and reuse activities by monitoring the condition of products at every stage of their use. The introduction of IoT also makes it possible to reduce information asymmetries, prevent delays and ensure a smooth flow of data between all stages of the green supply chain, from sourcing to production, consumption to recycling (Sarkis et al. 2021).

The benefits of IoT are multidimensional. Implementing this technology increases operational efficiency and productivity, reduces costs and creates new revenue streams. Process automation enables precise planning of logistics activities, resulting in improved customer service and reduced operational costs. Research shows a rapid increase in IoT adoption in the e-commerce sector, highlighting its growing importance as a tool for building competitive advantage (Khrais et al. 2023).

In reverse logistics, IoT plays an important role in solving the challenges of the randomness of returns, the unknown condition of reclaimed products and the imbalance between demand and returns. Implementing the technology as early as the production stage allows detailed data to be captured, improving the efficiency of remanufacturing and recycling processes. IoT offers application flexibility and the possibility to create dedicated information systems that support companies in integrating sustainable practices into their daily operations (Wang et al. 2023).

Paperless and blockchain

In logistics, we deal with a lot of information. Most of it is transmitted electronically between computer systems. However, even in the recent past, a large proportion of documents (e.g. invoices, waybills) were, and still are, transmitted in paper form in many companies. Reducing the printing of these documents has been helped by the paperless concept, which involves the digitalization of documentation processes in logistics. It saves time and operational costs, while contributing to environmental protection by reducing paper consumption. It also streamlines and speeds up communication in the supply chain (Masa'deh et al. 2024). A key element of implementation is the gradual implementation of new systems and appropriate staff training to effectively manage change and minimize resistance to transformation.

Simultaneously, blockchain technology is becoming an indispensable element of the digital transformation of logistics, offering transparency, security, and trust in supply chain management processes. Blockchain, a decentralized digital ledger technology, enables secure and transparent data sharing among all network participants (Sarkis et al. 2021). It allows the tracking of a product's lifecycle from production to the end consumer, significantly enhancing the credibility and efficiency of logistics operations. In practice, this means that every component in the supply chain-from raw materials through production, transportation, and distribution-can be accurately monitored in real time, reducing the risk of errors and data manipulation.

Blockchain technology also supports the digital transformation toward sustainable development and a circular economy. It enables companies to better manage resource flows, reduce excess, and efficiently share resources through peer-to-peer networks (Masa'deh et al. 2024). An example of such an application is decentralized energy management, allowing for the distribution of renewable resources and local system payments. Blockchain also aids in monitoring pollution emissions and waste generated during production processes. Such data, recorded transparently and made publicly accessible, empower consumers to make informed purchasing decisions and enable governments to manage environmental policies more effectively.

The benefits of implementing blockchain in logistics also include streamlining transactional processes, speeding up payments, and enhancing the reliability and transparency of data transfers (Sarkis et al., 2021). This enables more efficient management of complex operations, such as tracking products "from farm to table" in the food industry or monitoring materials and waste to assess companies' environmental performance (Wang et al. 2023). Blockchain also forms the foundation of smart contracts, which automate and secure the execution of agreements between parties, eliminating the need for intermediaries (Masa'deh et al. 2024).

The adoption of blockchain and paperless solutions holds particular significance in the context of Industry 4.0 and the digitalization of the logistics sector. A notable example of effective implementation is JD.com, which leverages IoT, AI, cloud computing, and blockchain to manage logistics and drive innovation. The company has also established over 70 data centers worldwide, supporting global logistics operations in line with digital transformation efforts (Yunlin 2023).

Paperless solutions and blockchain not only improve operational efficiency but also support sustainable development and help build a competitive advantage in the rapidly evolving logistics landscape. Together, they serve as the foundation of modern supply chains that address contemporary challenges such as environmental protection, the need for transparency, and rapid technological advancement.

Artificial intelligence

One of the main pillars of sustainable development in e-commerce logistics is the optimization of delivery processes through AI. By analyzing vast datasets, AI helps companies plan the most efficient delivery routes, reducing fuel consumption and CO_2 emissions. Additionally, AI enables demand forecasting, allowing businesses to manage inventory more effectively and minimize overproduction, which often leads to resource waste.

One of the biggest logistics challenges is optimizing delivery routes, especially in the context of the last mile. This stage is often the most environmentally burdensome, as it generates significant exhaust emissions and contributes to urban traffic congestion. Modern technologies like AI can process large amounts of data on traffic patterns, weather conditions, and fuel usage to determine optimal delivery routes. With dynamic route optimization, systems can continuously adjust delivery plans to avoid delays and unnecessary trips. For instance, Amazon has implemented a system that uses real-time data analysis to design routes for delivery drivers, minimizing fuel consumption. Similarly, UPS uses its ORION route optimization software to help drivers avoid left turns, reducing delivery times and cutting CO_2 emissions. This system dynamically optimizes individual pickup and delivery routes throughout the day based on current traffic conditions.

The implementation of AI systems enables companies to proactively mitigate risks and disruptions in global supply chains. For example, AI systems can extrapolate data, identify potential sources of delays, and suggest alternative solutions to prevent such issues (Cardinali and De Giovanni 2022). AI can also be used to predict future demand for products and transportation. For example, Alibaba Group has introduced an integrated online and offline retail model that allows customers to order items for home delivery or other locations, with a promise of delivery within 20 min. To meet these commitments, Alibaba has developed software specifically for scheduling vehicles and planning the shortest routes, efficiently utilizing available resources in the area to provide the fastest service. In logistics, robotic systems are increasingly used in distribution centers. These systems require minimal storage space and operate efficiently, contributing to streamlined operations and further enhancing sustainability (Wang et al. 2023).

Quantum technology

Quantum technology is based on the principles of quantum mechanics. It enables the simultaneous processing of vast amounts of data and the execution of complex computations that surpass the capabilities of classical computers. Quantum computing is regarded as a revolutionary tool that can significantly accelerate the development of innovative solutions across various fields.

The application of quantum technology is highly promising in logistics, where it can support process optimization, resource minimization, and the advancement of sustainable practices. With its ability to analyze massive datasets and model complex processes simultaneously, quantum computing introduces a new level of efficiency in supply chain management.

One key application of this technology is dynamic route and schedule planning. Traditional systems often face limitations in analyzing multidimensional route models that account for numerous variables such as distance, time, costs, and environmental conditions. Quantum technology allows for the parallel processing of multiple scenarios, enabling the selection of the most efficient routes. This results in savings in fuel, time, and operational costs while promoting sustainable development by reducing carbon emissions and the environmental impact of transportation (Sarkis et al. 2021).

Quantum technology also opens new opportunities in resource management. Its precision in modeling and measurement enables companies to monitor energy usage and other resources throughout the supply chain accurately. Organizations can adjust their processes in real time to minimize waste and increase efficiency. Furthermore, this technology supports the design of advanced environmental models, which can predict climate changes and evaluate the effectiveness of carbon capture technologies (Sarkis et al. 2021).

Another promising application of quantum computing is its support for circular economy initiatives. The optimization of recycling, refurbishing, and material recovery processes is made possible by advanced quantum algorithms, which maximize the utility of materials. This allows companies to design more sustainable products while enhancing the efficiency of operations aimed at closing the loop.

In a broader context, quantum technology supports global efforts to reduce carbon emissions and transition to renewable energy sources. For instance, sharing economy platforms leverage advanced technologies such as big data and cloud computing to optimize resource allocation and reduce vehicle idling. Quantum computing can further enhance these processes by analyzing data on an unprecedented scale, enabling more precise matching of supply and demand (Xie and Wang 2023).

As the demand for technology in the e-commerce sector grows, the importance of advanced solutions for data storage, scalability, and system availability also increases. Quantum computing, with its unique properties, offers companies the ability to maximize IT resources, improve data center efficiency, and provide greater operational flexibility. Strategies such as consolidation and virtualization, supported by quantum technology, allow businesses to achieve near-continuous service availability while reducing operational costs.

6 Paradox of Digitalization's Impact on Sustainable e-Commerce Logistics

As demonstrated in this chapter, digitalization plays a pivotal role in the development of e-commerce logistics, offering numerous benefits such as improved energy efficiency, waste reduction, and optimized resource utilization. Digital technologies like robotics, predictive analytics, big data, blockchain, IoT and AI help minimize waste, optimize processes, support circular economy initiatives, and enable companies to achieve corporate social responsibility (CSR) goals.

However, digitalization also introduces challenges that may conflict with the principles of sustainable development, creating a paradox (Sarkis et al. 2021; Xie and

Wang 2023). The growth of data centers, essential for supporting digital infrastructure, has led to a significant increase in energy consumption (Cardinali and De Giovanni 2022). The share of CO_2 emissions generated by these technologies rose from 0.8% in 2008 to 5.53% in 2018, with projections suggesting this figure reached 6.31% by 2020 (Xie and Wang 2023).

As Liu and De Giovanni (2019) noted, while technologies can reduce emissions in production processes, their widespread adoption increases resource consumption, as growing consumer demand drives the need for new products. Additionally, the literature highlights the risks of excessive optimism regarding digitalization's impact on sustainability (Sarkis et al. 2021). This is particularly evident with emerging technologies like blockchain and quantum computing, which have high implementation and operational costs and significant environmental impacts during their deployment phase.

Robotics systems, although effective at reducing waste and improving resource management, can lead to social challenges, such as the displacement of human labor by machines. Furthermore, the anticipated improvements in workplace safety and health outcomes due to digitalization are considered relatively modest While digital innovations in logistics address safety and health challenges, they often encounter resistance as they are perceived as threats to workers. This duality underscores the need to balance technological advancement with efforts to foster worker acceptance and engagement (Adeniran et al. 2024).

Researchers also caution that digitalization can exacerbate existing inequalities if the benefits of technological advancements are not equitably distributed (Kwilinski et al. 2023). De Giovanni (2022) introduced the concept of "responsible digitalization" emphasizing the need to balance the benefits of adopting new technologies with their negative social and environmental consequences (Cardinali and De Giovanni 2022). Examples of such actions include supporting the development of renewable energy sources to power data centers or designing more energy-efficient information systems (Kwilinski et al. 2023).

The paradox of digitalization in e-commerce logistics reveals that even technologies designed to promote sustainability can contribute to environmental issues. A "rebound effect" often occurs, where cost reductions lead to increased consumption and ultimately higher emissions (Cardinali and De Giovanni 2022).

To mitigate these challenges, a responsible approach to digital transformation is essential. This approach must include a comprehensive analysis of environmental and societal impacts. Only through such balanced measures can the full potential of digital technologies be realized without compromising the principles of sustainable development.

7 Regulations

Sustainable logistics in the e-commerce sector requires close collaboration between companies, governments, social organizations, and consumers. Regulations play a pivotal role in this process, promoting and enforcing eco-friendly and socially responsible practices. Their primary aim is to direct business activities toward reducing greenhouse gas emissions, minimizing waste, and investing in environmentally friendly technologies.

Environmental regulations encompass various mechanisms, including emission fees, taxation systems, and emissions trading schemes. These measures not only impose financial burdens on high-emission companies but also incentivize the development and adoption of innovative ecological solutions.

Studies (Wang et al. 2022; Wu and Tham 2023) indicate that environmental regulations can lead to sustainable competitiveness for companies through the adoption of advanced technologies. Similarly, Xie and Wang (2023) argue that implementing such regulations yields additional benefits, such as encouraging the use of better technologies and fostering long-term competitiveness. The threat of financial penalties, reduced subsidies, or reputational damage also motivates firms to take proactive steps toward sustainable development.

A key element of regulations is transparency policies, which require companies to disclose the environmental impact of their activities, thereby increasing transparency and enabling more effective monitoring of progress toward ecological goals. The growth of the digital economy further supports the enforcement of environmental regulations by enabling precise measurement of emissions and corporate activities. By leveraging digital tools, the effectiveness of such regulations is enhanced, as it becomes easier to assess corporate actions and their environmental impact.

In addition to formal mechanisms like fees and emissions trading systems, shaping consumer behavior is crucial. Raising awareness about responsible purchasing practices-such as thoughtful buying decisions, opting for consolidated deliveries, or reducing excessive packaging-plays a key role in supporting the achievement of sustainable development goals.

8 Conclusions

The digitization of logistics in the e-commerce sector is one of the key pillars of contemporary economic transformation. The implementation of advanced technologies such as IoT, AI, blockchain, and quantum computing brings numerous benefits to both businesses and the environment. Optimizing delivery routes, improving warehouse management, and reducing packaging waste are just a few examples of how digital technologies support sustainable development in e-commerce. By leveraging real-time data analysis, companies can better adapt their operations to changing market conditions while minimizing their environmental impact.

These technologies also enhance energy efficiency by enabling inventory management in models like just-in-time, eliminating overproduction, and reducing operational costs. Moreover, digitization supports the concept of a circular economy, which plays an increasingly significant role in building more sustainable logistics. A notable example is reverse logistics, which facilitates the recycling of products and their reintegration into the economic cycle.

However, this process comes with several challenges that demand attention. The growing demand for energy, the need for cooling data centers, and the widespread adoption of technologies with high carbon footprints can lead to a paradox: technologies designed to support sustainability may themselves become sources of additional environmental burdens. Furthermore, the advancement of digitization could contribute to social exclusion through job automation and the reduced role of humans in operational processes. These challenges call for a responsible approach that considers both technological benefits and potential negative consequences.

To effectively harness the potential of digitization, collaboration among multiple stakeholders is essential. Governments are introducing appropriate regulations to support sustainable development, such as carbon taxes and requirements for reporting environmental impacts. Companies should strive to implement innovations based on renewable energy and promote ecological initiatives such as zero-emission transport fleets and intelligent resource management systems. Consumers, on the other hand, can support these efforts through conscious choices, such as opting for consolidated deliveries, reducing impulsive purchases and returns, and favoring products manufactured according to sustainable principles.

In conclusion, the digitization of e-commerce logistics offers immense potential for creating a more sustainable and efficient sector. However, the success of this transformation depends on the ability to balance technological advancements with responsible management of their environmental and societal impacts. Further research into mitigating the negative effects of digitization and the implementation of policies promoting the responsible use of new technologies will be crucial. In this way, the e-commerce sector can become not only an economic driver but also a model for sustainable development in the digital age.

The article has several research limitations, such as focus on selected technologies and regions, and lack of empirical research. An attempt to address these may be also a further research direction.

References

Adeniran AO, Sidiq OB, Oyeniran GT, Adenira AA (2024) Sustainability impact of digital transformation in E-Commerce logistics. Int J Innov Mark Elem 4(1)

Al Mashalah H, Hassini E, Gunasekaran A, Bhatt D (2022) The impact of digital transformation on supply chains through e-commerce: Literature review and a conceptual framework. Transp Res Part E: Logist Transp Rev 165:102837

Bishop JDK (2022) Decarbonising transport with intelligent mobility. In: Intelligent Decarbonisation: Can Artificial Intelligence and Cyber-Physical Systems Help Achieve Climate Mitigation Targets. Springer: Cham, Switzerland, Vol 86, pp 163–172

Cardinali PG, De Giovanni P (2022) Responsible digitalization through digital technologies and green practices. Corp Soc Responsib Environ Manag 29(4)

Jurburg D, López A, Carli I, Chong M, De Oliveira LK, Dablanc L, De Sousa PR (2023) Understanding the challenges facing decarbonization in the E-Commerce logistics sector in latin america. Sustainability 15(22):15718

Kawa A, Pierański B (2021) Green logistics in e-commerce. LogForum 17(2):183–192

Khan SAR, Zhang Y, Nathaniel S (2020) Green supply chain performance and environmental sustainability: A panel study. LogForum 16(1):141–159

Khrais LT, Zorgui M, Aboalsamh HM (2023) Harvesting the digital green: A deeper look at the sustainable revolution brought by next-generation IoT in E-Commerce. Period Eng Nat Sci 11(6):5–13

Kong Z, Kong L, Bai K (2024) Research on the development path of E-commerce logistics in the context of carbon neutrality. In J Phys: Conf Ser 2735(1):012001. IOP Publishing

Kwilinski A, Lyulyov O, Pimonenko T (2023) Environmental sustainability within attaining sustainable development goals: The role of digitalization and the transport sector. Sustainability 15(14):11282

Lange S, Santarius T (2020) Smart green world?: Making digitalization work for sustainability. Routledge

Liu B, De Giovanni P (2019) Green process innovation through Industry 4.0 technologies and supply chain coordination. Ann Oper Res 1–36

Mangiaracina R, Marchet G, Perotti S, Tumino A (2015) A review of the environmental implications of B2C e-commerce: a logistics perspective. Int J Phys Distrib Logist Manag 45(6):565–591

Masa'deh RE, JaberM, Sharabati AAA, Nasereddin AY, Marei A (2024) The blockchain effect on courier supply chains digitalization and its contribution to industry 4.0 within the circular economy. Sustainability 16(16):7218

Masood MG, Alam J (2020) Green computing, technology efficiency and environmental sustainability in e-commerce-A solutions framework perspective

Mulholland E, Teter J, Cazzola P, McDonald Z, Gallachóir BPÓ (2018) The long haul towards decarbonising road freight–A global assessment to 2050. Appl Energy 216:678–693

Munoz F, Holsapple CW, Sasidharan S (2023) E-commerce. Springer Handbook of Automation. Springer International Publishing, Cham, pp 1411–1430

Nham NTH (2022) Making the circular economy digital or the digital economy circular? Empirical evidence from the European region. Technol Soc 70:102023

Wyman O (2021) Is E-commerce Good For Europe? Economic and environmental impact study. https://www.oliverwyman.com/content/dam/oliver-wyman/v2/publications/2021/apr/is-ecommerce-good-for-europe.pdf

Ortega A, Haq G, Tsakalidis A (2023) Carsharing in Europe: A critical review of policy, research, innovation, and practice. Transp Plan Technol 46(4):381–406

Sarkis J, Kouhizadeh M, Zhu QS (2021) Digitalization and the greening of supply chains. Ind Manag Data Syst 121(1):65–85

Shi S, Wang Y, Chen X, Zhang Q (2020) Conceptualization of omnichannel customer experience and its impact on shopping intention: A mixed-method approach. Int J Inf Manage 50:325–336

Štofejová L, Král' Š, Fedorko R, Bačík R, Tomášová M (2023) Sustainability and consumer behavior in electronic commerce. Sustainability 15(22):15902

Sun M, Grondys K, Hajiyev N, Zhukov P (2021) Improving the e-commerce business model in a sustainable environment. Sustainability 13(22):12667

Taniguchi E, Thompson RG, Yamada T (2016) New opportunities and challenges for city logistics. Transp Res Procedia 12:5–13

Tsakalidis A, Gkoumas K, Pekár F (2020) Digital Transformation supporting transport decarbonisation: Technological developments in EU-funded research and innovation. Sustainability 12:3762

Wang J, Dong X, Dong K (2022) How digital industries affect China's carbon emissions? Analysis of the direct and indirect structural effects. Technol Soc 68:101911

Wang Y, Yang Y, Qin Z, Yang Y, Li J (2023) A literature review on the application of digital technology in achieving green supply chain management. Sustainability 15(11):8564

Wu Y, Tham J (2023) The impact of environmental regulation, Environment, Social and Government Performance, and technological innovation on enterprise resilience under a green recovery. Heliyon 9(10)

Xie X, Wang J (2023) Study on the impact of industrial digitization on carbon emissions: evidence from China's Logistics Industry. Environ Res Commun 5(10):105008

Yunlin C (2023) Awareness of green logistics technology, certification, and standards by logistics practitioners at Chinese e-commerce company, Jing Dong. Asian J Shipp Logist 39(4):37–46

Innovative Analysis of Surface Layer Parameters Using Nondestructive Testing for Sustainable Product Lifecycle Management

Yaroslav Kusyi, Olha Kostiuk, Michal Hatala, and Vitalii Ivanov

Abstract The formation of regulated parameters of machine parts is ensured using their mechanical machining by rational interconnections of the technological system: machine tool—fixture—cutting tool—workpiece. The technological inheritability of material properties is an important criterion for analysis the effectiveness of technological process planning for product manufacturing. The concept of sustainable machining and Industry 4.0/5.0 requirements using functionally-oriented technologies requires new approaches to achieve the optimal quality parameters of machine parts. So, the technique of estimating the product's damageability has been developed using the LM-hardness method, which uses indicators of the homogeneity criterion of parts material. Experimental investigations on the influence of cutting modes and tool parameters on the characteristics of material homogeneity of the surface layers of aluminium workpieces during the different types of milling by end milling cutters of Sandvik Coromant 6–12 mm on vertical milling center HAAS MINIMILL were carried out. The developed technique allows for determining rational modes of cutting, taking into account the parameters of the tools to ensure the parameters of the quality of the products and economic indicators of the technological processes planning. Priority directions for further investigations are suggested.

Keywords Sustainable machining · Sustainable manufacturing · Technological inheritability · Process innovation

Y. Kusyi (✉) · O. Kostiuk
Lviv Polytechnic National University, Lviv, Ukraine
e-mail: yaroslav.m.kusyi@lpnu.ua

O. Kostiuk
e-mail: olha.s.kostiuk@lpnu.ua

Y. Kusyi · M. Hatala · V. Ivanov
Technical University of Kosice, Presov, Slovak Republic
e-mail: michal.hatala@tuke.sk

V. Ivanov
e-mail: ivanov@tmvi.sumdu.edu.ua

V. Ivanov
Sumy State University, Sumy, Ukraine

A. Kolinski and M. Adamczak (eds.), *Digitalisation of the Greening Supply Chain*, EcoProduction, https://doi.org/10.1007/978-3-031-88918-9_11

1 Introduction and Major Challenges

The diversity and differences in current operating conditions of different machines and their elements require changing priorities in the technological processes planning of machine parts according to the concepts of sustainable machining and Industry 4.0/5.0 requirements (Davim 2018; Liao et al. 2021).

Increasing the requirements for the quality parameters of engineering products has a need for a transition from object-oriented technologies to functionally-oriented and advanced technologies that include sustainable manufacturing, additive technologies, etc. (Fig. 1) (Cocca et al. 2018; Davim 2018). In this regard, the study of technological inheritance, which includes technological inheritability as a process and technological heredity as a property, is closely related to the search for new approaches to solving engineering and technical problems and the development of progressive methods in the technological processes planning at products manufacture (Kusyi and Kuk 2020; Kusyi et al. 2022). On the other hand, Al-Si alloys are ideal raw materials for replacing cast iron and other materials due to their mechanical properties, wear resistance, and physical characteristics (e.g., fluidity, density, thermal expansion coefficient, etc.) in mechanical engineering, aerospace, and automotive industries, etc. (Hernandez et al. 2017).

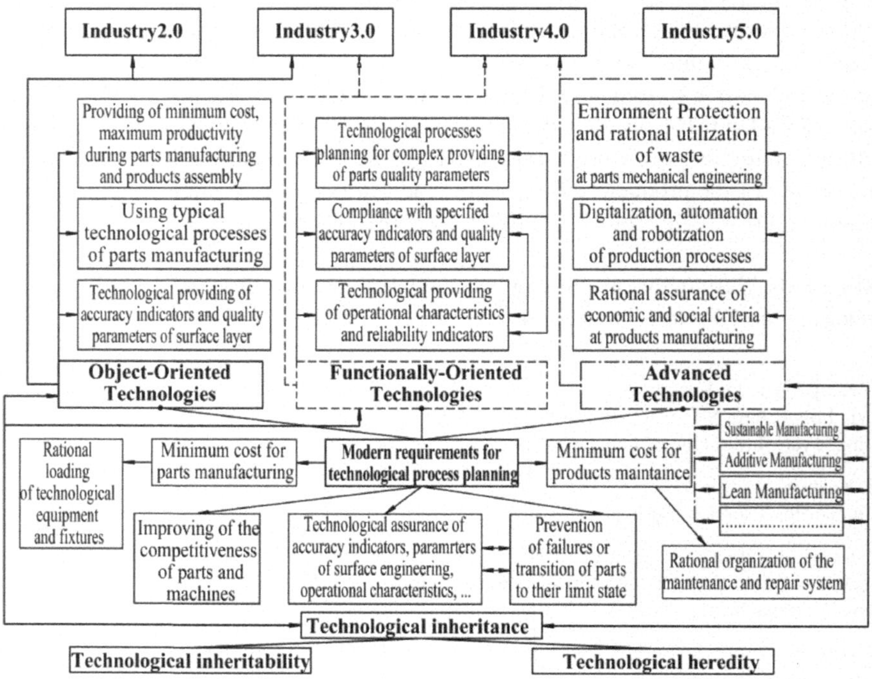

Fig. 1 Role of technological inheritance in mechanical engineering, own elaboration

Technological assurance of product quality is implemented by conventional approaches by the necessary current requirements: development of the optimal technological routes for surface machining at products manufacturing (Kusyi et al. 2022; Stupnytskyy 2020); detailed elaborating of finishing and finishing-strengthening technological operations for parts manufacturing (Aftanaziv et al. 2019; Kusyj and Kuk 2015); impact consideration of technological media elements for specific operational conditions of parts (Davim 2020; La Monaca et al. 2021); detailed investigations of the technological inheritability of material properties and surface parameters during products manufacturing (Demminger et al. 2016; Wang et al. 2022).

Meanwhile, the important laws of self-organization of technical systems (Haken 2006) and the requirements of Industry 4.0/5.0 (Chvykov et al. 2021) need to be taken into account under current technological processes planning for the rational machining and manufacture of mechanical engineering parts.

Industry 4.0 defines a process approach to the purposeful organization and management of a sequence of actions and works, the integration of all actions (operations) by changing the center of gravity from the vital function to process, which ensures the unity of management, improving organizational culture and allows for the rational implementation of PLM (Product/Part Life Cycles Management) technologies (Cocca et al. 2018; Hamrol et al. 2019; Latinovic et al. 2019; Liu and Xu 2017).

On the other hand, the main aim of Industry 5.0 is to synergize human capacities, leading-edge technologies, and adequate environmental protection. Industry 5.0 aims to bring social, economic, environmental, and human aspects back into product manufacture. The Industry 5.0 strategy is closely associated with the main ideas of sustainable machining and sustainable manufacturing (Longo et al. 2020; Nahavandi 2019).

In this regard, predicting the product's behaviour during important phases and stages in their life cycles, on the one hand, and development of effective approaches for analyses of product materials during their manufacture, on the other ones, according to the requirements of Industry 4.0/5.0 are the important tasks of current manufacturing engineering.

2 Literature Review

Nowadays, the sustainable machining of parts and the sustainable manufacture of products are closely related to the requirements of Industry 4.0/5.0, which ensure industrial growth, environmental protection, and economic and social criteria (Davim 2018; Davim 2020; Hamrol et al. 2019; Kumar et al. 2021). On the other hand, product quality is assured at all phases and stages in their life cycles (Ji and Abdoli 2023; Mauersberger et al. 2024). In this regard, total quality management, as all activities and necessary tasks to maintain a desired level of excellence (Psarommatis et al. 2020; Psarommatis and Bravos 2022), requires the analysis of specified processes and specific relationships between important design (Demminger et al. 2016; Liu and

Xu 2017), technological (Bembenek et al. 2022; Dobrotvorskiy et al. 2020; Kusyi and Kuk 2020) operational parameters (Kozłowski et al. 2023; Vlasiy et al. 2017), and reliability metrics (Dhillon 2017; Kusyi et al. 2022) in product life cycles.

The introduction of advanced chipless machining of mechanical engineering parts, in particular, generative technologies (3D printing technologies, coating, surfacing, etc.) along with the production of high-precision blanks, reduces the share of machining by cutting in the manufacture of products, which requires the improvement of not only milling, turning, grinding, etc. (Hernandez et al. 2017; La Monaca et al. 2021; Seneci et al. 2024), but also the modernization of methods and means of technical control of regulated parameters of material and surfaces of products (Antosz et al. 2025; Kozłowski et al. 2024; Muzyka et al. 2020).

Therefore, the development of effective methods for nondestructive testing of the regulated parameters of products' surface layers and their materials is a progressive direction in the development of modern manufacturing engineering.

Comprehensive investigations of product manufacturing require the consideration of hereditary phenomena in their life cycles during their machining, assembly, and further operation (Davim 2018; Kozłowski et al. 2023). So, in that regard, technological inheritance is determined by a generalized indicator for analysing a mechanical engineering product's technological inheritability (process) and technological heredity (characteristics). Investigation of changes due to the transfer of properties of workpieces from previous technological operations (in the technological process) and technological steps (in the technological operations) to the following during their machining by cutting or abrasive machining, heat pressure treatment (surface plastic deformation), etc. is called technological inheritability in mechanical engineering. Technological inheritability directly provides the formation conditions of accuracy indicators, machine parts' surface quality, operational properties, and reliability indicators. Preserving the specific properties in machine parts is called "technological heredity" (Demminger et al. 2016; Kusyi et al. 2024).

At the current phase of mechanical engineering development, most research is concerned with technological inheritability during the machining and assembly of products. Based on a synergistic approach, due to the self-organization of technological systems at parts manufacture and the assembly of machines, the increase of applicable (positive) and elimination of harmful (adverse) properties of products takes place. The process of forming technological parameters and material properties of parts is ensured (Davim 2018; Kusyi et al. 2022). The important terms "technological object", "technological system", and "technological media" are used during the estimation, analysis, and assessment of the technological inheritability of product properties (Kusyi et al. 2022, 2024). A technological object is any of the objects. As a result of their interaction, the state of the production object is determined or changed at machine manufacture and its components. A combination of technological objects that engage with each other at the parts production and assembly of the machine implements a technological system. Technological systems of operations, processes, production units, enterprises, etc., are distinguished during machine parts manufacturing (Demminger et al. 2016; La Monaca et al. 2021).

The technological process of product manufacture is identical to the transformation (inheritability) and preservation (heredity) of the properties of the object's production during the interaction of the relevant technological media with the production object. Therefore, the product properties are determined by the properties of the workpiece and the technological media that formed them, which are the carriers of the mechanism of inheritability and transformation of the properties of the product in the process of their manufacture (Aftanaziv et al. 2019; Kusyi and Kuk 2020). The technological system's structure at the technological operation level is the technological system: "machine tool—fixture—cutting tool—workpiece" (Davim 2018). The technological media at workpiece machining is directly formed by the metal-cutting tool and the fixture and indirectly by the metal-cutting machine (Kusyi et al. 2022, 2024). Machined material and functional surfaces of the parts are the carriers of inherited information. They actively participate in the technological process, passing through various technological operations and steps and testing the actions of determining physical factors (Kusyi and Kuk 2020; Kusyi et al. 2022).

Nowadays, the influence of technological system elements on the technological inheritability of the parameters of parts from aluminium alloy needs further thorough research. Aluminium alloys are widely used in various branches of industry due to their high specific strength and corrosion resistance (Hernandez et al. 2017; Salerno et al. 2024). However, the broader use of these light alloys for manufacturing machines parts and equipment is restrained by their low hardness and easily damaged surface. The optimal choice of materials (Dobrotvorskiy et al. 2020) and the geometric shape of products (Vlasiy et al. 2017) are used to increase the wear resistance of aluminium alloys. In addition, metallic chromium coatings (Dubei et al. 2022; Shatskyi et al. 2023a, b; Peltier and Thierry 2022), non-metallic oxide coatings formed by anodizing (Student et al. 2021), and plasma electrolytic oxidation (Ropyak et al. 2021, 2023) are used to solve technological problems. The behaviour of parts describes with cracks (Shats'kyi et al. 2019) and under load (Shatskyi et al. 2023a, b). The authors (Bembenek et al. 2022) suggested using image processing methods to detect damage to the surface of products.

Sustainable machining of mechanical engineering parts contains the estimation and advanced analysis of the relationships in social, economic, and environmental dimensions for technological assurance of quality parameters of products according to Industry 4.0/5.0 requirements (see Fig. 2) (Davim 2020; Hamrol et al. 2019). Social, environmental, and economic challenges are key elements of sustainable manufacturing. These problems are related to ensuring the sustainable development of the economy, social well-being, and environmental protection. The environmental direction has improved its important role in the sustainable machining of products. The priority problems in this field are closely associated with the rational use of different types of industrial energy in productive processes, minimizing the consumption of cutting fluids, recycling chips and cutting tools, etc. (Fig. 2) (Davim 2018; Kumar et al. 2021; Ropyak et al. 2023).

On the other hand, the economic criteria determine the competitiveness of products and the effectiveness of implementing advanced technologies for producing mechanical engineering parts and assembling machines (Cocca et al. 2018). The main

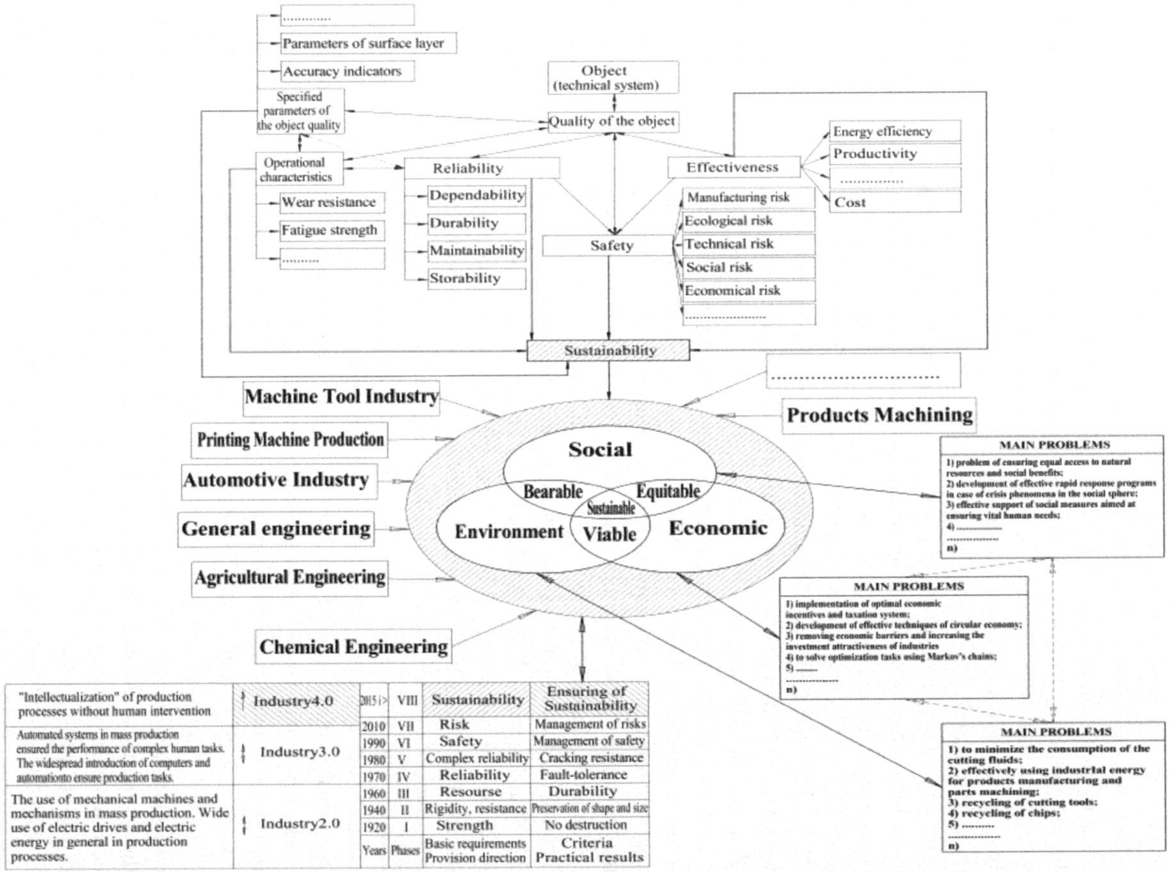

Fig. 2 Relationships between the main directions in sustainable machining, own elaboration

problems in economic directions are strongly connected with the implementation of optimal economic incentives and taxation system, the development of effective techniques of circular economy, removing economic barriers and increasing the investment attractiveness of industries, the development of techniques to solve optimization tasks using the technique of Markov's chains, etc. (Fig. 2) (Bochko et al. 2022; Podvalna et al. 2021; Kusyi et al. 2022). The social criteria are related to ensuring equal access to natural resources and social benefits, development of effective rapid response programs in case of crisis phenomena in the social sphere, adequate support of social measures aimed at ensuring vital human needs, etc. (Fig. 2). A characteristic feature of current manufacturing engineering is the comprehensive assurance of technological, economic, ecological, and social aspects of sustainable manufacture and machining in integrated and digitalized design and process preparation of parts production (Davim 2018; Nahavandi 2019).

Mechanical engineering production is provided by object-oriented and functionally-oriented technologies to manufacture machine parts (Kusyi and Kuk 2020). Object-oriented technologies are described as the step-by-step implementation of interconnected stages using the typical technological processes of parts manufacturing. The minimum cost of product manufacture is an optimization criterion for choosing an optimal variant of the specific technological process (Stupnytskyy 2020). Functional analysis according to the operational characteristics of parts and machines is not carried out for object-oriented technologies, relying on the designer's production experience, high qualifications, and technical literacy (Kusyi et al. 2022; Stupnytskyy 2020).

Technical conditions for products, as a rule, do not regulate the entire combination of basic parameters of surface engineering. Still, they are often limited to surface macrohardness or microhardness values and roughness. In addition, technological parameters and indicators are not always considered during parts manufacture. As a rule, they are chosen from high productivity provision, accurate loading of technological equipment, and material properties that change under the influence of operational indicators (high temperatures, power loads, the influence of aggressive environments, etc.). Mechanical engineering parts with different operational characteristics and reliability indicators are obtained for specified operating conditions during various technological processes of their manufacture (Davim 2018; Liao et al. 2021).

Parameters of technological processes (structure and sequence of technological operations, cutting modes, methods of parameter control, etc.) significantly affect the generation of technological defects at the stages of product development and their future evolution at the operational stage. Each type of technological influence used by the elements of the technological media, machine tool—fixture—cutting tool—workpiece is characterized by typical types of defects associated with certain process sequence violations or an unfavorable combination of factors (La Monaca et al. 2021). At the same time, functionally-oriented technologies are characterized by the technological assurance of the most influential and necessary operational characteristics of the product in compliance with the accuracy and quality parameters of the surface layer of the product that the designer normalizes (Kusyi et al. 2022; Stupnytskyy 2020).

The general technological chain is analysed in the Life Cycles of Parts/Products to take into account the principles of technological inheritability of their properties. The decision-making algorithm for choosing a rational structure and parameters of the technological process is implemented to consider the operation conditions of certain parts in the technical system and technological support of their reliability indicators and operational characteristics. At the same time, the technological inheritability of product properties is analysed from rough to finishing and finishing-strengthening technological operations of product machining (Kusyi and Kuk 2020). The result of research using the technological inheritability of quality parameters of products is the analysis of the dynamic changes in the homogeneity of the material structure after technological machining at individual steps by the developed surface treatment routes to ensure the regulated quality parameters of parts (Kusyi et al. 2022, 2024).

The analysis of changes in material homogeneity ensured by different methods at the stage of blanks production and operation of products (Davim 2018; La Monaca et al. 2021) should also be implemented for the surfaces of products at their manufacture. This makes it possible to check the correctness of the construction of technological routes for machining the surfaces of mechanical engineering parts by the technical conditions.

The main goal of this scientific work is to research of the influence of technological system's elements: machine tool—fixture—cutting tool—workpiece on the technological inheritability of the parameters of the part after aluminium alloy machining.

3 Research Methodology

The development of continuous damage mechanics allows us to analyse the reasons for the decreasing physical and mechanical properties of parts materials (Murakami 2012). In this regard, increasing the reliability of the physical and mechanical characteristics determined for the surface layer of their materials is ensured by a new approach to materials science research using the LM-hardness method. The LM-hardness method was developed under the guidance of academician Lebedev more than 20 years ago. The main idea of the LM-hardness method is to analyse the homogeneity of the surface layer of parts materials after mechanical treatment, on the one hand, or its damageability, on the other hand. In this regard, we analysed the state of the surface layers of parts materials after cutting using the scattering of the measured and processed parameters of their physical and mechanical indicators, in particular macrohardness, carried out by the same devices under the same conditions (Lebedev et al. 2006; Muzyka et al. 2020).

Homogeneity serves as a parameter that integrally characterizes the current state of the material after its machining that is defined by the hardness control results and characterized by the Weibull distribution (Lebedev et al. 2006):

$$P(\sigma) = 1 - e^{-(\sigma/k)^m} \tag{1}$$

where P(s)—Calculated probability; s—dispersion; k, m– parameters of the Weibull distribution.

A structurally sensitive parameter, namely: Weibull homogeneity coefficient (m) characterizes a level of structure degradation of product material, m ∈ [0; ∞]. It is determined by (Lebedev et al. 2006; Muzyka et al. 2020):

$$m = d(n)/(2.30259 \cdot S(\ell g(H))) \tag{2}$$

$$S(\ell g(H)) = \sqrt{\frac{1}{n-1} \cdot \sum_{i=1}^{n} \left(\ell g(H_i) - \overline{\ell g(H)}\right)^2}; \ \overline{\ell g(H)} = \frac{1}{n} \cdot \sum_{i=1}^{n} \ell g(H_i) \tag{3}$$

where d(n) is a parameter that depends on the number of experimental measurements n of macrohardness H_i.

Studies have established that higher values of the coefficient m correspond to a lower level of dispersion of hardness characteristics and, obviously, a better organization of the structure and a low level of material damageability. On the contrary, a lower value of the Weibull homogeneity coefficient (m) corresponds to a higher level of material damageability and a worse organization of the structure of the product material. The number of tests is 25–50 measurements of the macrohardness of the surface layer, which assures regulated accuracy in assessing the degree of homogeneity of the product material (Lebedev et al. 2006; Muzyka et al. 2020).

The coefficient of variation also describes the degree of dispersion of the researched physical and mechanical characteristics, in particular, macrohardness (Lebedev et al. 2006; Muzyka et al. 2020):

$$\nu = \sqrt{\frac{1}{n-1} \cdot \sum_{i=1}^{n} \left(H_i - \overline{H}\right)^2} \Big/ \overline{H}, \tag{4}$$

where $\overline{H}$ describes the average value of macrohardness.

The criterion of material homogeneity is established based on theoretical and experimental research. It is defined by material homogeneity as a parameter for the estimation and analysis of the technological inheritability of the indicators of the machined surfaces. The following parameters describe it: the Weibull homogeneity coefficient (m), the material constant A_m, the technological damageability D and its growth rate j_D, and refinements in the technological steps of machining of machine parts e_i (Fig. 3) (Kusyi et al. 2024).

The material constant, the increase of which determines the tendency of the material structure to homogeneity, and its decrease indicates the tendency of the material to damageability, is determined by (Kusyi and Kuk 2020; Kusyi et al. 2024):

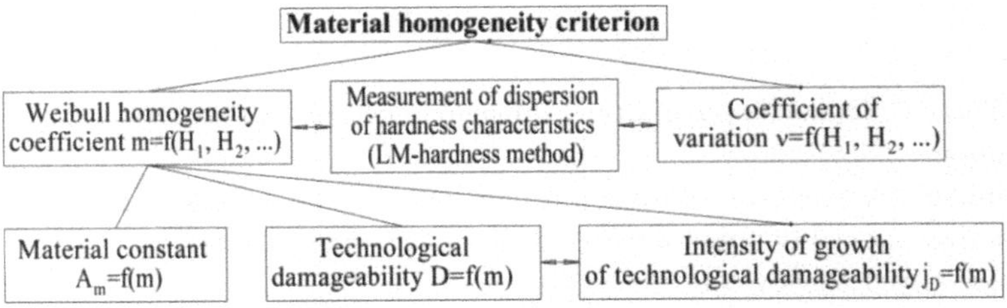

Fig. 3 The structure of the material homogeneity criterion, based on (Kusyi et al. 2024)

$$A_m = \left((m + 2)/(4m + 4)\right)^{1/m} \quad (5)$$

The dispersion of hardness characteristics describes the change in technological damageability D according to Brinell, Rockwell, and Vickers during the Weibull homogeneity coefficient (m) changes (Kusyi and Kuk 2020; Kusyi et al. 2024):

$$D = (m_{matr.} - m_i)/m_{matr.}, \quad (6)$$

where $m_{matr.}$, m_i is the Weibull homogeneity coefficient (m), characterizing the dispersion of the base material's hardness and the product's i-th section, respectively.

The intensity of growth of technological damageability j_D is calculated by (Kusyi et al. 2024):

$$j_D = (m_{matr.} - m_i)/m_i \quad (7)$$

An increase in the current values of technological damageability and the intensity of its growth indicates the susceptibility of the material to damage. On the contrary, a decrease in the value of technological damageability and the intensity of its growth determines the tendency to homogeneity of the material structure (Kusyi et al. 2024).

In this regard, the material homogeneity criterion (MHC) for the analysis of the technological inheritability of the regulated quality parameters of parts during their production by cutting, surface plastic deformation, and technical control or testing of the material properties of the surface layer by the LM-hardness method, which is determined by the combination of indicators for each technological step/operation, is realized by (Kusyi et al. 2024).

- for specific technological steps/operations during the technological inheritability of material properties of the product

$$\begin{aligned} \mathrm{MHC} \in [\mathrm{m} = \mathrm{f}(\mathrm{H_1 H_2} \ldots\ldots); \nu = f(\mathrm{H_1}, \mathrm{H_2} \ldots\ldots); \\ \mathrm{A} = \mathrm{f}(\mathrm{m}); \mathrm{D} = f(\mathrm{m}); \mathrm{j_D} = \mathrm{f}(\mathrm{m}); \varepsilon_i = \mathrm{m_i}/\mathrm{m_{i-1}} \end{aligned} \quad (8)$$

- for the technological inheritability of quality parameters of the product

$$\left|\begin{array}{l} m_{x_1} < \ldots m_{x_k} < \ldots < m_{x_n} \to \infty;\ \nu_{x_1} > \ldots \nu_{x_k} > \ldots > \nu_{x_n} \to 0 \\ A_{x_1} < \ldots A_{x_k} < \ldots < A_{x_n} \to 1;\ D_{x_1} > \ldots D_{x_k} > \ldots > D_{x_n} \to 0; \\ \varepsilon_{x_1} > \ldots \varepsilon_{x_k} > \ldots > \varepsilon_{x_n} \to 0;\ \prod_{i=1}^{n} \varepsilon_i \geq \varepsilon_p \end{array}\right| \text{for} \quad T_0 = \sum_{k=1}^{n} t_{ok}, \tag{9}$$

- for the state of the structurally heterogeneous material of the surface layer of the product

$$m < 10;\ \nu \to 1(100\,\%);\ A_m < 0.878;\ D \to 1, \tag{10}$$

- for the state of the structurally homogeneous material of the surface layer of the product

$$m \geq 10;\ \nu \to 0;\ A_m \geq 0.878;\ D \to 0, \tag{11}$$

where t_{0k} is the direct manufacturing or cutting time during the performance of the k-th technological step (machining method); T_0 is the total direct manufacturing or cutting time for machining the specified part surface.

The material homogeneity criterion consists of six indicators: Weibull homogeneity coefficient (m), coefficient of variation n, material constant A_m, technological damageability D, intensity of growth of technological damageability j_D, and calculated refinement e_i (see Eq. (8)). The dynamic changes of WHC indicators from the first (1) to (n) technological operation is represented by Eq. (9). Conditions (10)–(11) determine the general trends towards the state of structurally heterogeneous (10) and structurally homogeneous (11) material of the product.

The value of the Weibull homogeneity coefficient (m) in Eqs. (8)–(11) is determined by (2and3), coefficient of variation—by (4), material constants A_m—by (5), technological damageability D and the intensity of its growth j_D—by (6) and (7), respectively.

The number of indicators (9)–(12) is used to analyse the material homogeneity of the blanks and workpieces at the specified technological step/operation and is defined by the complexity of the performed technological tasks for different design phases. The tasks' complexity increases from initial (rough) operations to finishing and finishing-hardening operations. In particular, the material homogeneity of the blank is analysed by indicators or m (A_m), and indicators m. A_m, D, j_D, etc. analyse the material homogeneity of the final part.

4 Experimental Research, Analysis of the Obtained Results

Castings in sand-clay molds with dimensions of 165’155’20 mm were used for experimental research. The castings were produced of aluminium alloy AK21M2.5H2.5 (Table 1).

The surface of the casting is machined by milling end cutters. “Sandvik” end cutters for rough and semi-rough milling Ø 6–12 mm were used for experimental research. The experimental samples were machined on a vertical milling center HAAS MINIMILL CNC (Fig. 4a).

Two series of research studies were carried out on a variety of experimental samples machining. Initially, the determined surfaces of the experimental sample were machined by rough milling, alternately using four end cutters for semi-rough milling. In the next stage, the specified surfaces of the sample are machined by semi-rough milling using four end cutters for rough milling. For real production, rough milling cutters are used for preliminary (rough) machining of the workpiece, and semi-rough milling cutters are used for semi-rough one.

The cutting modes (milling width B, cutting depth t, feed per tooth Sz, spindle speed n, cutting speed V) are selected by the recommendations of “Sandvik” for the specified technological equipment: for Ø 6 mm: B = 3.0 mm, t = 6.0 mm, Sz = 0.1 mm/tooth, n = 3100 min^{-1}, $V_{cut.}$ = 58.43 m/min; for Ø 8 mm: B = 3.0 mm, t = 8.0 mm, Sz = 0.1 mm/tooth, n = 4000 min^{-1}, $V_{cut.}$ = 100.53 m/min; for Ø 10 mm: B = 3.0 mm, t = 10.0 mm, Sz = 0.1 mm/tooth, n = 5800 min^{-1}, $V_{cut.}$ = 182.21 m/min; for Ø 12 mm: B = 3.0 mm, t = 12.0 mm, Sz = 0.1 mm/tooth, n = 5800 min^{-1}, $V_{cut.}$ = 218.66 m/min (Fig. 4b).

Machining was carried out during one outspanning of the workpiece for one step during machining using end cutters. Control of the machined surface of the casting was carried out after each technological step on the TP5006 device according to the standard method (Fig. 4c).

The metal-cutting tool is the weakest link among the elements of the technological media. Therefore, research the influence of structural and geometrical parameters and types of metal-cutting tools on the formation of quality indicators of the machined surfaces of products is an essential task of current manufacturing engineering. MHC indicators: Weibull homogeneity coefficients (m) after each treatment are calculated by (2)–(3), material constants A_m—by (5), technological damageability D—by (6), intensity of growth of technological damageability j_D—by (7) in the Mathcad 15 environment based on research results.

Graphic dependences of the effect of varieties and design parameters of the end cutters on indicators of the criterion of surface homogeneity are shown in Figs. 5, 6 and 7.

An increase of the cutting speed from $V_{cut.}$ = 58.43 m/min to $V_{cut.}$ = 218.65 m/min during machining by end cutters for semi-rough milling (Fig. 5) using the same feed per tooth and cutting width is characterized by a decrease in the Weibull homogeneity coefficient (m) and an increase in the material damageability after rough milling. After semi-rough milling, use the same range of $V_{cut.}$ the Weibull homogeneity

Table 1 Chemical composition of the material AK21M2.5H2.5

Chemical element	Fe	Si	Mn	Ni	Cr	Ti	Al	Cu	Pb	Mg	Zn	Sn
Mass fraction, %	0 to 0.91	20.5	0.2 –0.4	2.2–2.8	0.2–0.4	0.1–0.3	74.9–69.5	2.2–3.0	0 to 0.04	0.2–0.5	0 to 0.2	0 to 0.01

source own elaboration

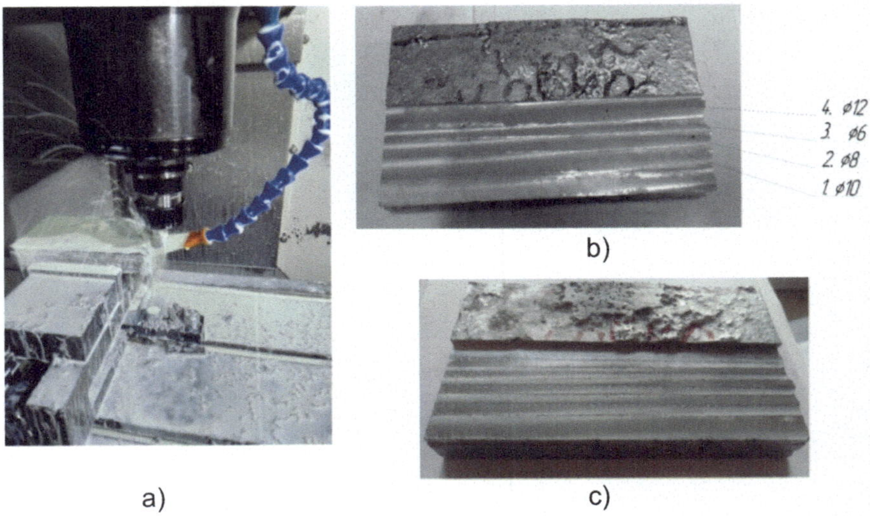

Fig. 4 Machining of the workpiece using end cutters Ø 6 mm, Ø 8 mm, Ø 10 mm, Ø 12 mm **a**, the surface of the experimental sample after machining **b** and surf of hardness measurements using the LM-hardness method, own elaboration

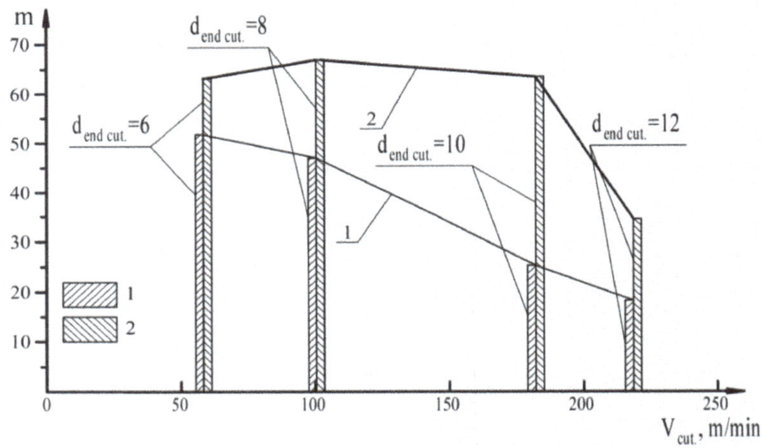

Fig. 5 The dynamics of changes in the Weibull homogeneity coefficient (m) depend on the cutting speed of $V_{cut.}$, after cutting: 1—after rough milling; 2—after semi-rough milling using "Sandvik" end cutters, own elaboration

coefficient (m) increases by 1.22–2.51 times, and the indicator changes provide direction for an increase in the end cutter diameter of 6–10 mm. For Ø 6–10 mm, the homogeneity coefficient stabilizes after semi-rough milling, which is explained by cutting the defective material layer after the first milling. However, for Ø 12 mm of the dynamics of change, the Weibull homogeneity coefficient (m) decreases. Compared

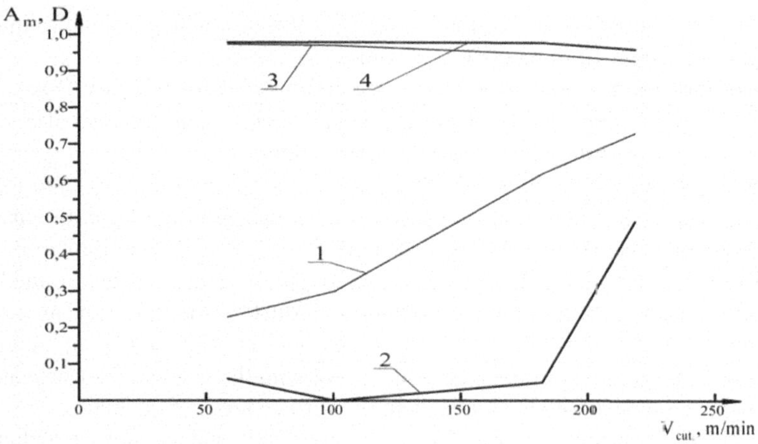

Fig. 6 The dynamics of changes in technological damageability D depend on the cutting speed of $V_{cut.}$ after cutting: 1—after rough milling; 2—after semi-rough milling and material constant A_m: 3—after rough milling; 4—after semi-rough milling using "Sandvik" end cutters, own elaboration

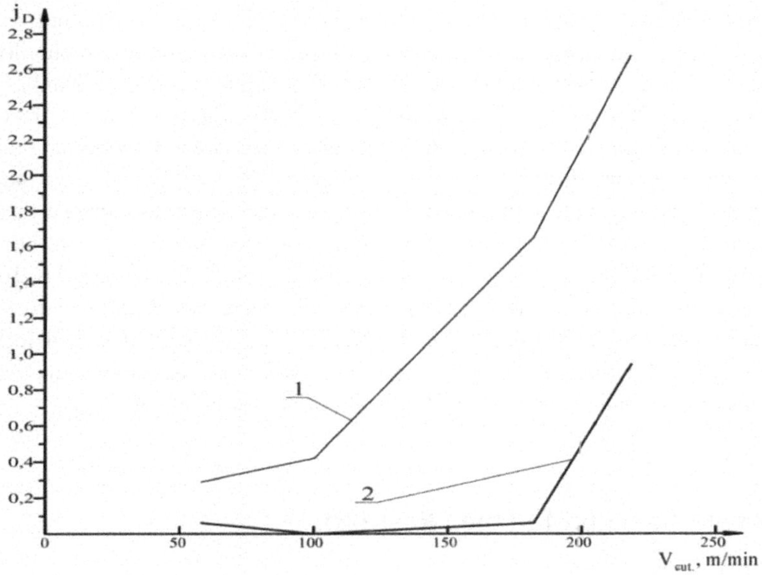

Fig. 7 Dynamics of changes of damageability growth j_D as a function of the cutting speed $V_{cut.}$ after cutting: 1—after rough milling; 2—after semi-rough milling using "Sandvik" end cutters, own elaboration

with rough milling, the value (m) increases by 1.9 times, which is smaller than the same value for the Ø 10 mm cutter (2.51 times).

The surface layer was analysed using technological damageability D as a characteristic of MHC because the measured values were stabilized after the second machining in three of the four standard sizes of cutters.

Figure 6 shows the dynamics of changes in material constants A_m and technological damageability D during rough and semi-rough milling of aluminium alloy blanks—an increase of the cutting speed from $V_{cut.} = 58.43$ m/min to $V_{cut.} = 218.65$ m/min for rough milling using semi-rough end cutters is determined by a decrease in the value of the material constant A_m from 0.974 to 0.929 and an increase in technological damageability D by 3.17 times. This is explained by an increase in the specific contact area of the metal-cutting tool with the machined surface and the presence of many technological defects in the surface layer of the blank.

The material constant stabilizes ($A_m = 0.979$–0.980), and technological damageability D decreases to almost zero after semi-rough milling using rough end cutters Ø 6–10 mm at the same values of $V_{cu.}$. At the same time, semi-rough milling using end cutters Ø 12 is accompanied by a decrease in technological damageability D by 1.4 times by increasing the material constant A_m from 0.929 to 0.96, which is determined by the decrease in technological defects after preliminary rough machining.

Figure 7 shows the dynamics of changes in the intensity of growth of technological damageability j_D after rough and semi-rough milling aluminium alloy blanks.

An increase of the cutting speed from $V_{cut.} = 58.43$ m/min to $V_{cut.} = 218.65$ m/min for rough milling using semi-rough end cutters is characterized by an increase of the intensity of damage ability j_D by 9.24 times, which is explained by an increase in the specific contact area of the metal-cutting tool with the workpiece surface and the presence of a significant number of structural imperfections of the material in the surface layer of the blank. After semi-rough milling, rough end cutters Ø 6–10 mm were used at the same values as $V_{cut.}$ the parameter j_D decreases almost to zero. At the same time, semi-rough milling using end cutters Ø 12 is accompanied by a decrease in the intensity of damageability j_D by 2.85 times, which is ensured by a decrease in the number of technological defects after preliminary rough machining.

5 Conclusions and Further Research

The following main conclusions have been drawn based on the results.

1. A metal-cutting tool is the weakest link of the technological media; therefore, researching its influence on the formation of quality parameters of the machined surfaces of products is an essential task of current mechanical engineering.
2. The influence of cutting modes on the characteristics of the material homogeneity for the surface layer of the experimental sample from aluminium alloy was investigated in the process of rough milling using end cutters for semi-rough milling and semi-rough milling using end cutters for rough milling of the company

"Sandvik" Ø 6–12 mm on vertical milling center HAAS MINIMILL CNC. An increase of the cutting speed from $V_{cut.} = 58.43$ m/min to $V_{cut.} = 218.65$ m/min during cutting using end cutters for semi-rough milling at the same feed per tooth and cutting width is characterized by a decrease of the Weibull homogeneity coefficient (m), a decrease of the constant material A_m from 0.974 to 0.929 and an increase of technological damageability D by 3.17 times. After semi-rough milling at the same values as $V_{cut.}$, the Weibull homogeneity coefficient (m) increases by 1.22–2.51 times, the material constant stabilizes (A_m = 0.979–0.980), and the technological damageability D decreases almost to zero. For Ø 6–10 mm, the homogeneity coefficient stabilizes after semi-rough milling, which is explained by cutting the defective material layer after the first treatment. However, for Ø 12, mm the Weibull homogeneity coefficient (m) decreases. Compared to the first treatment, the value (m) increases by 1.9 times, which is smaller than the same indicator for the cutter Ø 10 mm (2.51 times). At the same time, semi-rough milling using end cutters Ø 12 is accompanied by a decrease of technological damageability D by 1.49 times, an increase of the material constant A_m from 0.929 to 0.96, which is determined by a decrease in the number of technological defects after preliminary rough machining.
3. Further research should be carried out for a broader nomenclature of machine parts and materials to research the influence of elements of the technological system: machine tool—fixture—cutting tool—workpiece (part) on the technological inheritability. The obtained results will be introduced into the current mechanical engineering production practice.

Acknowledgements This work was funded by the Slovak Research and Development Agency (APVV-21-0228) and the Ministry of Education, Science, Research and Sport of the Slovak Republic (VEGA 1/0391/22 and KEGA 017TUKE-4/2023), The NAWA Ulam Programme (grant number BPN/ULM/2022/1/00045), NextGenerationEU through the Recovery and Resilience Plan for Slovakia under project No. 09I03-03-V01-00094, the International Association for Technological Development and Innovations.

References

Antosz K, Kozłowski E, Sęp J, Prucnal S (2025) Application of machine learning to the prediction of surface roughness in the milling process based on sensor signals. Materials 18(1):148

Aftanaziv I, Shevchuk L, Samsin I, Strutynska L, Strogan O (2019) Development of a technology for the surface strengthening barrel channels in the large-caliber artillery guns. East-Eur J Enterp Technol 3(1–99):11–18

Bembenek M, Mandziy T, Ivasenko I, Berehulyak O, Vorobel R, Slobodyan Z, Ropyak L (2022) Multiclass Level-Set segmentation of rust and coating damages in images of metal structures. Sensors 22(19):7600

Chvykov P, Berrueta TA, Vardhan A, Savoie W, Samland A, Murphey TD, Wiesenfeld K, Goldman DI, England JL (2021) Low rattling: A predictive principle for self-organization in active collectives. Science 371(6524):90–95

Cocca P, Marciano F, Rossi D, Alberti M (2018) Business software offer for industry 4.0: The SAP case. IFAC-PapersOnLine, 51(11):1200–1205

Davim JP (Ed) (2018) Sustainable Machining. Materials Forming, Machining and Tribology. Springer; Softcover reprint of the original 1st ed. 2017 edition

Davim JP (2020) Manufacturing in digital industries: Prospects for Industry 4.0 (Advanced Mechanical Engineering), Book 6. Walter de Gruyter Gmbh, Berlin/Boston, Germany/USA

Demminger C, Mozgova I, Quirico M, Uhlich F, Denkena B, Lachmayer R, Nyhuis P (2016) The consept concept of technical inheritance in operation: Analysis of the information flow in the life cycle of smart product. Procedia Technol 26:79–88

Dhillon BS (ed) (2017) Engineering Systems Reliability, Safety, and Maintenance. An Integrated Approach, 1st edn. CRC Press, Boca Raton

Dobrotvorskiy S, Balog M, Basova Y, Dobrovolska L, Zinchenko A (2020) Concept of the software for materials selection using. Net technologies. In: Advanced Manufacturing Processes. Inter-Partner 2019. Tonkonogyi V, Ivanov V, Trojanowska J, Oborskyi G, Edl M, Kuric I, Pavlenko I, Dasic P (Eds) Part of the Lecture Notes in Mechanical Engineering book series (LNME), Springer: Cham, Switzerland

Dubei OY, Tutko TF, Ropyak LY, Shovkoplias MV (2022) Development of analytical model of threaded connection of tubular parts of Chrome-Plated metal structures. Metallofiz Noveishie Tekhnol 44(2):251–272

Haken H (2006) Information and Self–Organization. A Macroscopic Approach to Complex Systems: 3rd edned. Springer, Berlin

Hamrol A, Gawlik J, Sładek J (2019) Mechanical engineering in industry 4.0. Manag Prod Eng Rev 10(3):14–28

Hernandez FCR, Ramírez JMH, Mackay R (2017) Al-Si Alloys: Automotive, Aeronautical, and Aerospace Applications. Cham, Switzerland, Springer

Ji X, Abdoli S (2023) Challenges and opportunities in product life cycle management in the context of industry 4.0. Procedia CIRP 119:29–34

Kozłowski E, Antosz K, Sęp J, Prucnal S (2024) Integrating sensor systems and signal processing for sustainable production: Analysis of cutting tool condition. Electronics (Switzerland) 13(1):185

Kozłowski E, Borucka A, Oleszczuk P, Jałowiec T (2023) Evaluation of the maintenance system readiness using the semi-Markov model taking into account hidden factors. Eksploatacja i Niezawodnosc—Maintenance and Reliability 25(4)

Kumar K, Zindani D, Davim JP (2021) Sustainable Manufacturing and Design. Woodhead Publishing, Cambridge, United Kingdom

Kusyi Y, Kuk A (2020) Investigation of the technological damageability of castings at the stage of design and technological preparation of the machine Life Cycle. IOP Conf Ser: J Phy 1426:012034

Kusyi Y, Stupnytskyy V, Kostiuk O, Onysko O, Dragašius E, Baskutis S, Chatys R (2024) Control of the parameters of the surface layer of steel parts during their processing applying the material homogeneity criterion. Eksploatacja i Niezawodnosc—Maintenance and Reliability 26(3)

Kusyi Y, Stupnytskyy V, Onysko O, Dragašius E, Baskutis S, Chatys R (2022) Optimization synthesis of technological parameters during manufacturing of the parts. Eksploatacja i Niezawodnosc—Maintenance and Reliability 24(4):655–667

Kusyj J, Kuk A (2015) A centrifugal vibration strengthening method devised to improve technological reliability of machine parts. East-Eur J Enterp Technol 1(7):41–51

La Monaca A, Murray JW, Liao Z, Speidel A, Robles-Linares JA, Axinte A, Hardy MC, Clare AT (2021) Surface integrity in metal machining—Part II: Functional performance. Int J Mach Tools Manuf 164:103718

Latinovic T, Preradović D, Barz CR, Pop Vadean A, Todić M (2019) Big data as the basis for the innovative development strategy of the Industry 4.0. IOP Conference Series: Materials Science and Engineering, 477(1):012045

Lebedev AA, Makovetskii IV, Muzyka NR, Volchek NL, Shvets VP (2006) Assessment of damage level in materials by the scatter of elastic characteristics and static strength. Strength Mater 38(2):109–116. https://link.springer.com/article/https://doi.org/10.1007/s11223-006-0022-9.

Liao Z, la Monaca A, Murray J, Speidel A, Ushmaev D, Clare A, Axinte D, M'Saoubi R (2021) Surface integrity in metal machining—Part I: Fundamentals of surface characteristics and formation mechanisms. Int J Mach Tools Manuf 162:103687

Liu C, Xu X (2017) Cyber-physical machine tool—the era of machine tool 4.0. Procedia CIRP 63:70–75

Longo F, Padovano A, Umbrella S (2020) Value-oriented and ethical technology engineering in industry 5.0: A human-centric perspective for the design of the factory of the future. Appl Sci 10(12):4182

Mauersberger A, Reinhardt H, Alaluss M, Mahdaviasl M, Ihlenfeldt S (2024) Overview of methodologies for the lifecycle management of smart product-service systems. Procedia CIRP 128:174–179

Murakami S (2012) Continuum Damage Mechanics—A Continuum Mechanics Approach to the Analysis of Damage and Fracture. Springer, Dordrecht/Heidelberg/London/New York

Muzyka N, Shvets V, Boiko A (2020) Procedure and instruments for the material damage assessment by the LM-Hardness method on the In-Service scratching of structure element surfaces. Strength Mater 52:432–439

Nahavandi S (2019) Industry 5.0—a human-centric solution. Sustainability 11(16):4371

Peltier F, Thierry D (2022) Review of Cr-Free coatings for the corrosion protection of aluminum aerospace alloys. Coatings 12(4):518

Podvalna H, Bochko O, Kuziak V, Stasyuk N (2021) Economic efficiency of road transport vehicles of fleet and its impact on commercial indices and production plan of a motor transport enterprise. E3S Web of Conferences 234, 00056

Psarommatis F, Bravos G (2022) A holistic approach for achieving Sustainable manufacturing using zero defect manufacturing: A conceptual framework. Procedia CIRP 107:107–112

Psarommatis F, May G, Dreyfus P-A, Kiritsis D (2020) Zero defect manufacturing: state-of- the-art review, shortcomings and future directions in research. Int J Prod Res 58(1) 1–17

Ropyak L, Shihab T, Velychkovych A, Bilinskyi V, Malinin V, Romaniv M (2023) Optimization of Plasma Electrolytic Oxidation technological parameters of deformed aluminum alloy D16T in flowing electrolyte. Ceramics 6(1):146–167

Ropyak L, Shihab T, Velychkovych A, Dubei O, Tutko T, Bilinskyi V (2023) Design of a Two-Layer Al–Al_2O_3 coating with an Oxide layer formed by the Plasma Electrolytic oxidation of Al for the corrosion and wear protections of steel. Prog Phys Metals 24(2):319–365

Ropyak LY, Velychkovych AS, Vytvytskyi VS, Shovkoplias MV (2021) Analytical study of "crosshead—Slide rail" wear effect on pump rod stress state. J Phys: Conf Ser 1741(1), art. no. 012039

Salerno E, Casotti D, Gualtieri E, Ballestrazzi A, Gazzadi GC, Bolelli G, Lusvarghi L, Rota A, Valeri S, Paolicelli G (2024) Friction and wear characteristics of DLC-terminated coatings deposited on AlSi10Mg alloy produced by Additive Manufacturing. Surf Coat Technol 494(2):131422

Seneci G, Pancera PV, Cappellini C, Abeni A, Attanasio A (2024) An innovative method to model run-out phenomena in micro-milling by using cutting force signal. Mater Res Proc 41:1867–1876

Shats'kyi IP, Makoviichuk MV, Shcherbii AB (2019) Influence of a flexible coating on the strength of a shallow cylindrical shell with longitudinal crack. J Math Sci (United States) 238(2):165–173

Shatskyi I, Makoviichuk M, Ropyak L, Velychkovych A (2023) Analytical model of deformation of a functionally graded ceramic coating under local load. Ceramics 6(3):1879–1893

Shatskyi IP, Makoviichuk MV, Ropyak LY (2023) Equilibrium of laminated Cu/Ni/Cr coating under local load. Nanosistemi, Nanomateriali, Nanotehnologii 21(2):379–389

Student MM, Pohrelyuk IM, Hvozdetskyi VM, Veselivska HH, Zadorozhna KR, Mardarevych RS, Dzioba YV (2021) Influence of the composition of Electrolyte for Hard Anodizing of aluminum on the characteristics of Oxide layer. Mater Sci 57(2):240–247

Stupnytskyy V (2020) Computer-aided conception for planning and researching of the functional-oriented manufacturing process. In.: Advanced Manufacturing Processes, part of the Lecture Notes in Mechanical Engineering, pp 309–320

Vlasiy O, Mazurenko V, Ropyak L, Rogal O (2017) Improving the aluminum drill pipes stability by optimizing the shape of protector thickening. East-Eur J Enterp Technol 1(7(85)):25–31

Wang DW, Zhou YH, Yao XY, Dong YP, Feng SH, Liu ZY, Wang H, Yan M (2022) Inheritance of microstructure and mechanical properties in laser powder bed fusion additive manufacturing: A feedstock perspective. Mater Sci Eng, A 832:142311

Yu BO, Maletska OI, Tsitska NE, Kapral OR (2022) Paradigm of a country competitiveness under conditions of digital economy. Rev Econ Financ 20:572–580

The manufacturer's authorised representative in the EU is Springer Nature Customer Service Centre GmbH, Europaplatz 3, 69115 Heidelberg, Germany. If you have any concerns regarding our products, please contact ProductSafety@springernature.com

Printed and bound by CPI Group (UK) Ltd, Croydon, CR0 4YY
30/09/2026
02225569-0005